Understanding
Vineyard Soils

Understanding Vineyard Soils

Second Edition

Robert E. White

OXFORD
UNIVERSITY PRESS

OXFORD
UNIVERSITY PRESS

Oxford University Press is a department of the University of
Oxford. It furthers the University's objective of excellence in research,
scholarship, and education by publishing worldwide.

Oxford   New York
Auckland   Cape Town   Dar es Salaam   Hong Kong   Karachi
Kuala Lumpur   Madrid   Melbourne   Mexico City   Nairobi
New Delhi   Shanghai   Taipei   Toronto

With offices in
Argentina   Austria   Brazil   Chile   Czech Republic   France   Greece
Guatemala   Hungary   Italy   Japan   Poland   Portugal   Singapore
South Korea   Switzerland   Thailand   Turkey   Ukraine   Vietnam

Oxford is a registered trademark of Oxford University Press
in the UK and certain other countries.

Published in the United States of America by
Oxford University Press
198 Madison Avenue, New York, NY 10016

© Oxford University Press 2015

Library of Congress Cataloging-in-Publication Data
White, R. E. (Robert Edwin), 1937–
Understanding vineyard soils / Robert E. White. — 2nd edition.
pages cm
Includes bibliographical references and index.
ISBN 978–0–19–934206–8 (alk. paper)
1. Grapes—Soils.   2. Viticulture.   3. Terroir.   I. Title.
S597.G68W45 2015
634.8—dc23
2014024000

9 8 7 6 5 4 3 2 1
Printed in the United States of America
on acid-free paper

Contents

Foreword

There is no branch of agriculture in which soils are more rhapsodized and venerated than in viticulture. It is easy to understand why this is so, just as it is easy to understand why our ancestors worshipped the sun and the moon, feared an eclipse, and ascribed meaning to thunder.

The aromas and flavors of wine are astonishingly diverse and at best are compellingly beautiful. Few processed agricultural products command a higher price than the world's finest wines. Since viticulture and winemaking alone are unable to replicate the quality of benchmark fine wines in secondary locations, it is evident that something in the environment in which the wine came into being governs its fundamental quality. What might that be?

The preferred candidate for most wine drinkers is the soil. It can be seen, touched, measured, dug, even smelled. Vines roots penetrate deeply into often poor, rocky soils. A vine rootstock spends its entire life (sometimes a century or more) in one place. In certain key European wine regions such as Burgundy, one winegrower making his wine in a uniform manner from the same grape variety will produce a very different wine from two vineyards 10 meters or less apart. It is hard, confronted with sensory evidence of this sort, and lifted by the emotion that great wine provokes, not to set about constructing a mythology of soil.

Enter Professor Robert White, soil scientist. In *Understanding Vineyard Soils*, he describes what soil is and how vineyard soils differ from one another. (Soils are not rocks and are much more than an accumulation of minerals.) We learn about the life forms in the soils and about its air spaces, structure, and profile.

The presence of water in the soil is of huge significance for vines: is it best to rely on what nature delivers, or is irrigation a wiser strategy? To what extent are vines nourished by the soil medium in which their roots are buried, and to what extent do they draw sustenance from light and air? What task do roots perform, and how do they do it? How does insect life in the soil favor and challenge viticulture? What, exactly, is a "healthy soil" for vines, and how might a winegrower set about creating one? What do organic and biodynamic approaches to vineyard husbandry bring, or neglect? How will climate change affect the world's vineyard soils?

This book will be of great practical value to anyone growing vines or looking after a vineyard, but it is also essential reading for those wine lovers who are prepared to demythologize their thinking about the soils in which vines grow. The soil doesn't explain the flavor of a wine, any more than thunder portends the wrath of the gods. A vine is a plant, not a person; it does not feed but photosynthesizes. A soil's structural properties, its biota and the water relations it offers to the plant, might be far more important than its precise mineral spectrum. Professor White tells us what can be known at present about vineyard soils (hence the importance of this revised edition), rather than what can be fabulated.

Aroma and flavor in wine are the product of a series of hugely complex equations. Soil is of great significance; so too are landforms, climate, weather, plant genetics, biochemistry, organic chemistry, and human cultural practices. This book will help you lay the groundwork for a deeper understanding of the complex issues that underlie a pleasure both simple and profound: drinking wine.

Andrew Jefford, journalist and author

Prades le Lez, France

Preface to the Second Edition

> What makes wine great is the interaction of the roots and the soil. The complexity, the character, the suppleness of the wine is reflected in that complex relationship.
>
> ▪ Quoted in *Daily Wine News*, April 19, 2010, and attributed to one of the fifth-generation winemakers at Perrin & Fils, owners of Chateau de Beaucastel in the Chateauneuf-du-Pape appellation, southern Rhone, France

I would like to start this preface by acknowledging many colleagues and friends who have assisted in various ways in making this second edition possible. First, I would like warmly to thank Dr. Mark Krstic, Dr. Robert Bramley, Dr. Tony Proffitt, and Dr. Ian Porter who read and commented on one or more chapters. In addition, I would like to thank all those who supplied material of one kind or another that has been reproduced here—Dr. Lilanga Balachandra, Dr. Tapas Biswas, Dr. Robert Bramley, Dr. Eduard Hoffman, Dr. Jonathan Holland, Dr. Mark Imhof, Professor Greg Jones, Mr. Brad Johnston, Mr. Richard Merry, Mr. Martin Peters, Dr. Tony Proffitt, Mr. Stuart Proud, Dr. Loothar Rahman, Dr. Belinda Rawnsley, and Dr. Melanie Weckert.

My research for *Understanding Vineyard Soils* and the earlier book *Soils for Fine Wines* has taken me to many exciting wine regions, where I have met numerous charming winegrowers, as well as been given the opportunity to sample some memorable wines. Winegrowing is truly one of the few completely vertically integrated agricultural industries, where consumers are aware of, and appreciate, the connection between the soil in which the vines grow and the seductive liquid in the glass. Although this revised edition retains the essentials of what I have learned from these experiences, it also updates knowledge about the soil–wine nexus that has evolved in the past five years.

In particular, chapter 1 of this edition expands on the concept of soil health and provides a broader picture of the diversity of soils on which grapevines grow around the world. Chapter 2 offers more practical details about site selection and soil preparation for vineyards. Chapters 3 and 4 on vine nutrition and water relations are the most challenging, more so for those with a less technical background. However, I have tried to simplify the important concepts without losing the underlying scientific rigor. Chapter 5 contains much new material on soil biology, the role of soil organic matter, and benchmarking soil biological properties. As before, chapter 6 draws together several themes bearing on a soil's "fitness for purpose" and updates the discussion of topical issues such as climate change; water supply and its effects on fruit quality; organic, biodynamic, and conventional viticulture; managing natural soil variability to meet specific winemaking objectives; and integrated production systems and sustainability. As before, my wife Annette has given me constant support on our many enjoyable travels to wine regions and during the exacting task of writing.

Robert E. White
Melbourne, April 2, 2014

Understanding
Vineyard Soils

1

What Makes a Healthy Soil?

Perceptions of Soil Health

Soil scientists used to speak of soil quality, a concept expressing a soil's "fitness for purpose." The prime purpose was for agriculture and the production of food and fiber. However, to the general public soil quality is a rather abstract concept and in recent years the term has been replaced by soil health. A significant reason for this change is that health is a concept that resonates with people in a personal sense. This change is epitomized in the motto "healthy soil = healthy food = healthy people" on the website of the Rodale Institute in Pennsylvania (http://rodaleinstitute.org/).

One consequence of this change is an increasing focus on the state of the soil's biology, or life in the soil, an emphasis that is expressed through the promotion of organic and biodynamic systems of farming. Viticulture and winemaking are at the forefront of this trend. For example, Jane Wilson (2008), a vigneron in the Mudgee region of New South Wales, is quoted as saying, "the only way to build soil and release a lot of the available minerals is by looking after the biology," and Steve Wratten (2009), professor of ecology at Lincoln University in New Zealand has said, "Organic viticulture rocks! It's the future, it really is." This exuberance has been taken up by Organic Winegrowers[1] New Zealand, founded only in 2007,

---

[1] The term "winegrowing" refers to the integrated process of growing grapes and making wine. Grapegrowers and winemakers are collectively called "winegrowers."

1

who have set a goal of "20 by 2020," that is, 20% of the country's vineyards under certified organic management by the year 2020.

The Cornell Soil Health Assessment provides a more balanced assessment of soil health (Gugino et al., 2009). The underlying concept is that soil health is an integral expression of a soil's chemical, physical, and biological attributes, which determine how well a soil provides various ecosystem functions, including nutrient cycling, supporting biodiversity, storing and filtering water, and maintaining resilience in the face of disturbance, both natural and anthropogenic. Although originally developed for crop land in the northeast United States, the Cornell soil health approach is readily adapted to viticulture, as explained by Schindelbeck and van Es (2011), and which is currently being attempted in Australia (Oliver et al., 2013; Riches et al., 2013).

A most important subtext of the Cornell approach is the recognition of "inherent" and "dynamic" factors of soil health. Collectively, the inherent factors determine a soil's "sense of place," being primarily the influence of geology (the rocks), climate, organisms living on and in the soil, topography (the relief), and time. In the absence of human intervention, these factors can be considered immutable during a human's life span, in contrast to the dynamic factors such as organic matter, soil pH, nutrient availability, structure, and water supply, which can be manipulated by human activities, often quite quickly. Thus to understand "what makes a soil healthy," we need to understand something about the advantages and disadvantages conferred by a soil's sense of place and how these are modified by interaction with dynamic factors that are subject to human influence.

The remainder of this chapter provides an introduction to the inherent factors of soil health, which leads logically in chapter 2 to criteria for the selection of vineyard sites and preparation of those sites for planting. In succeeding chapters, the main dynamic factors and how they can be managed to best advantage in a vineyard are discussed in more detail. The final chapter draws these various threads together to answer the question "what is a healthy soil?" in the context of the objectives of individual grape growers and winemakers.

Inherent Factors of Soil Health

Well before the current interest in soil health, a soil scientist named Hans Jenny (1941) published a book called *Factors of Soil Formation* in which he demonstrated with many examples how the soil that formed in a particular place was a product of the interaction of several soil forming factors. These factors are the same as those identified as inherent factors in the language of soil health. Table 1.1 gives a summary of the soil forming factors, their components, and some consequences of their action. To understand a little of the complexity of the interactions, we can follow the steps in soil formation when a rock is exposed to weathering for the first time.

Table 1.1 Summary of Soil Forming Factors and their Effects

Factors	Components	Characteristics and active agents	Features of soils formed
Parent material	*Parent material in place*		
	Igneous rocks—derived from volcanic activity	Acidic rocks, e.g., granite, rhyolite, rich in quartz and pale-colored Ca^a, K, and Na feldspars	Soils on granite have a high proportion of quartz grains, often coarse; well drained if deeply weathered; low fertility
		Basic rocks, e.g., basalt, dolerite, low in quartz and rich in dark-colored ferromagnesian minerals	Soils on basalt often high in clay and Fe and Al oxides; can be poorly drained unless old and well weathered; generally fertile
	Sedimentary rocks—weathered rock material deposited under water or by wind and subsequently compressed	Varying size of rock particles forming mudstones (small), sandstones, shales, and conglomerates (large); fine-grained chalk and limestone	Soils on mudstone often poorly drained; soils on sandstone and conglomerates of low fertility; soils on limestone vary with the content of impurities
	Metamorphic rocks—igneous and sedimentary rocks subjected to heat and pressure that has changed the original rock structure and minerals	Granite metamorphosed to gneiss, sandstone to quartzite, shale to schist, limestone to marble	Variable weathering and properties, depending on the original rock that was metamorphosed
	Transported parent material		
	Weathered rock materials moved from their original place	Water, gravity, ice and wind deposits; when compressed, can form sedimentary rocks	Alluvial (water) and colluvial soils (gravity), loess (wind); usually well drained and fertile
Climate	Macroclimate—the regional climate applicable over tens of kilometers	Modified by latitude and elevation, temperature affects the rate of organic matter decomposition and other biochemical reactions; moisture availability affects weathering, leaching, and clay movement down the soil profile	Current and past climates are reflected in present-day and buried ("fossil") soils through differences in depth, color, organic matter, structure, translocation of clay, organic compounds, salts, and oxides, and ultimately the profile form (see box 1.1)
	Mesoclimate—applicable to a subregion or site, such as a valley or slope		
	Microclimate—very local, as within the canopy of a vineyard		

(continued)

Table 1.1 (continued)

Factors	Components	Characteristics and active agents	Features of soils formed
Organisms	Plants Animals Humans	Plant species differ in the chemical properties of their leaf litter; roots and earthworms form channels; microorganisms decompose and humans add, organic residues	Under conifers and heath, Fe and Al organic complexes translocated to form podzols and podzolic soils; humified organic matter forms under deciduous trees and grassland
Topography	Slope Aspect	Water and gravity move material down steep slopes; solifluction occurs in thawed-over-frozen soil Slopes facing the equator are warmest	Soils at the top of slopes are shallow with good drainage; depth increases downslope but drainage usually is poorer, so Fe colors change from orange-red through yellow to blue-gray
Time	Tens to hundreds of years (recent volcanic eruptions, human activity) to tens of thousands of years (glacial and interglacial periods of the Pleistocene—box 1.2) to more than 1,000,000 years (in stable landscapes)	Modifies the action of the other factors	Soils range from shallow with little profile development to deep with well-developed profiles as time passes; younger soils are generally more fertile

[a] Chemical symbols for the elements are explained in chapter 3.

How Does Soil Form?

Even where grapevines appear to be growing in rock, we know that initially a soil, no matter how meager, was present because when a rock is first exposed to weathering, a soil begins to form. Figure 1.1 shows a very early stage of soil formation on granite rock.

Soil is distinguished from weathered fragments of rock by the presence of living organisms and organic matter. An assortment of inorganic minerals inherited from the parent rock forms the structural framework—solid material and spaces—within which lives a diversity of organisms. Life in the soil is a struggle, but the cycle of growth, death, and decay is essential for healthy soil function in cultivated fields, forests, grasslands, and vineyards. Organisms ranging in size from bacteria, invisible to the human eye, to fungi, insects, earthworms, and burrowing animals, feed on the residues of plants and other organisms and become themselves food for subsequent generations (see "The Soil Biomass," chapter 5). Provided the developing soil is not unduly disturbed, the processes of soil formation result in a growing depth of soil that often exhibits a characteristic profile. Box 1.1 presents a general description of a soil profile.

Although complex in terms of the internal processes of mineral weathering, percolating rainwater, exchange of gases, and the cycle of life, the formation of soil in the example of figure 1.1 tells us little about why soils are so variable in the

Figure 1.1 A thin mantle of soil forming on granite rock with early-colonizing plants. The scale is 10 cm.

Box 1.1 The Soil Profile

A vertical exposure of soil in a pit or road cutting is called a soil profile. Often there are obvious changes in color and composition from the surface (enriched with organic matter from plant litter) to the subsoil and parent material below, as can be seen, for example, in figure 1.8. Where these changes in color and composition are the result of soil formation in different parent materials, the profile is called "layered." Often a lower layer represents a buried or "fossil" soil.

Where the profile changes are due to different biophysical processes within the same parent material, the layers are called horizons, labeled A, B, and C from top to bottom. Horizons may be subdivided. For example, the upper zone of an A horizon that contains dark-brown organic matter is labeled A1, as distinct from the paler zone immediately below, labeled A2 (figure B1.1.1). The A and B horizons comprise the soil proper, with the C horizon being weathering parent material. An O horizon is a superficial organic horizon composed of partially decomposed plant litter.

Figure B1.1.1 A vineyard soil in the Gippsland region Victoria, Australia, showing distinct A and B horizons. The A horizon is a bleached sandy loam and the B horizon is a pale orange-mottled clay. There is a thin layer of organic enrichment at the top of the A horizon (A1). The major divisions in the scale are at 10 cm intervals. See color insert.

(continued)

Box 1.1 *(continued)*

On the basis of texture (the "feel" of the mix of soil particles), soil profiles may be classed as

- uniform—little change in texture with depth (e.g., figure 1.9)
- gradational—a gradual increase in clay content with depth (e.g., figure 1.10)
- duplex or texture-contrast—a change from a "light" textured A horizon (sandy loam to sandy clay loam) to a "heavy" textured B horizon (clay loam or clay), often with an sharp boundary between the two (e.g., figure B1.1.1)

The A horizon (topsoil) is easy to distinguish from the B horizon (subsoil) in a duplex profile, but this separation is less obvious in a gradational profile and even less so in a uniform profile. In the second and third cases, a distinction between A and B horizons can be made on the basis of organic matter content, which is greater in the topsoil and usually decreases to an insignificant amount below 20 cm or so.

landscape. The answer to this question lies in the multifaceted interaction of the soil forming factors that can vary on a scale of a few meters to tens of kilometers and the effect this variable interaction has on the resultant soil's physical, chemical, and biological properties. Some examples of the different possible outcomes for soil formation illustrate this point.

Rudimentary Soils on Slate and Schist

Where mountain building has been active, such as in the Alps of northern Italy, the original rocks that may have been igneous or sedimentary (see table 1.1) can become much folded and faulted. During this process, the rocks are metamorphosed such that, for example, a sedimentary mudstone becomes hardened and converted to a shale or schist. Although soil will form on surface outcrops of such rock, where the land slope is steep, the action of gravity and water causes the developing soil to be eroded and remain shallow. Figure 1.2 shows an example of one such rudimentary soil that is used for viticulture, formed on fractured shale in the Collio del Friuli region of northeast Italy. Similar soils formed on schist support vineyards in the Central Otago region, New Zealand, and the Coteaux du Languedoc region, France. Because of their shallowness these soils are not very fertile, but given their fractured parent rock and land slope they are usually well drained. Hence once the grapevines are established, with roots exploring the weathering rock, these soils can be excellent for producing fruit for wine that is rich in flavors and aromas. The distinctive Pinot Grigio wines from the Friuli region and Pinot Noir wines from the Central Otago region attest to this outcome.

Figure 1.2 A rudimentary soil with some root development on a fractured shale in the Friuli region of northeast Italy. The scale is 10 cm.

Soils on Limestone

Retired geologist John Davis (2010) of Pepper Tree winery in the Hunter Valley, Australia, has said, "Limestone is rare in Australia but it's a key to the greatness of the famous wine regions of France. When you look at why France is so fantastic at wine, it's underlain by limestone from one end to the other."

Although many would challenge Davis's view of the relationship between great wine and limestone, it is true that limestone is an abundant rock that underlies many wine regions, especially in Europe. Limestone is a sedimentary rock formed from precipitated calcium carbonate minerals (chemical composition, $CaCO_3$), or from the calcified skeletons of countless marine animals and algae deposited in ancient seabeds. The purest form of limestone is the Chalk formation of northwest Europe, which forms the parent material of soils in the renowned Champagne and Loire regions of France. Other limestones, such as in the Dolomitic Alps of northern Italy, have up to half the calcium substituted by magnesium (Mg) to form dolomite ($Ca,MgCO_3$). However, most limestones have a proportion of impurities such as clay, quartz, and iron oxide particles. As limestone weathers, the carbonate dissolves to release Ca^{2+} and carbonate

Figure 1.3 An organic-rich Brown Earth on limestone before cultivation in the Côte d'Or, Burgundy region, France. (White, 2003)

(CO_3^{2-}) ions[2] in solution, the latter breaking down to form carbon dioxide (CO_2), so that only the ions and insoluble impurities remain to form soil.

Given this mode of formation and weathering, it is no surprise that soils formed on limestone are usually shallow. One estimate for the rate of soil formation on Chalk in Britain is 1 cm of soil in 5,000 years. Normally, an organic-rich A horizon rests directly on weathering rock, as seen in shallow profile of a Rendzina on fractured limestone in the Côte d'Or of Burgundy region, France (figure 1.3). However, if the impurities in the limestone are substantial, or the soil is augmented by an external source of mineral matter, deeper soils form. An example of the latter is the Terra Rossa soil of the Coonawarra region in South Australia, where, during the glacial (cold) periods of the Pleistocene epoch when the sea level was lowered, strong winds blew silt-size particles from the coastline onto the exposed ridges of limestone. As a result a deeper, red soil profile formed over the porous limestone, which is capped by a hard layer of calcrete (figure 1.4). As with other limestone regions around the world (e.g., Burgundy, St. Emilion in France, Paso Robles in California), the wines produced from the Coonawarra Terra Rossa are distinctive and much prized, especially those made from Cabernet

[2] Ions and their charges are explained in chapter 3.

Figure 1.4 A Terra Rossa soil formed on calcrete-capped porous limestone in the Coonawarra region, Australia. Vines are in the background. See color insert.

Sauvignon fruit. The limited soil volume offered by these soils exerts a control on the vigor of the vines.

Box 1.2 gives a summary of the ages of rocks. Note that although the geological time scale extends back many millions of years, the present-day soil need not necessarily be very old because in most instances the parent rock has undergone many cycles of weathering, erosion, and deposition in an ever-changing landscape. Older soils will have been buried (see box 1.1), and current soil formation may have proceeded for only a few thousand years or less.

Soils on Transported Materials

As exposed rocks weather, some of the weathered material remains in place, but much can be transported elsewhere by gravity, water, ice, or wind. On steep slopes, especially in cold or very dry climates where vegetation is sparse, rock fragments tumble downslope under the influence of gravity to form colluvial deposits at the footslopes. Such colluvial deposits in the shadow of the Mayacamus Mountains of the Napa region, California, where they are called benches (e.g., Rutherford, Oakville, and St. Helena), comprise some of the best vineyard sites in the valley because of their low fertility, good drainage, and favorable local climate. Elsewhere, the steep slopes themselves are extensively used for vineyards as in the Valais and Vaud regions of Switzerland (figure 1.5)

Box 1.2 The Age of Rocks

Compared with the 6,000 years or so of recorded human history, the earth's age of 4.6 billion years is unimaginably long. Nevertheless, even events that occurred millions of years ago have relevance today because of the unending cycles of rock formation, weathering, erosion, and deposition that have laid the foundations of, and helped to shape, the present earth's surface. As the science of geology developed, the history of earth's rocks was divided into a time scale consisting of eras, periods, and epochs. Most information is available for rocks dating from the appearance of the first multicellular life forms about 650 million years ago, through many subsequent epochs to the present. Periods within eras are usually associated with sequences of sedimentary rocks that were deposited in the area now known as Europe and recognized there for the first time. Although examples of these rocks are found elsewhere, the European time divisions have generally been accepted worldwide, except in cases such as the Ediacaran period of the Pre-Cambrian era whose defining fossils were first identified in the Flinders Ranges of South Australia. Table B1.2.1 gives a simplified geological time scale from the Cambrian period, when life forms exploded in diversity and number, to the present.

Table B1.2.1 The Geological Time Scale

Era	Period	Epoch	Start time (million years ago)
Cainozoic ("young life")	Quaternary	Recent	0.011
		Pleistocene (the ice ages)	2
	Tertiary (the age of mammals)	Pliocene	5
		Miocene	23
		Oligocene	36
		Eocene	53
		Palaeocene	65
Mesozoic ("middle life" and the age of reptiles)	Cretaceous		145
	Jurassic		205
	Triassic		250
Palaeozoic ("old life")	Permian		290
	Carboniferous		360
	Devonian		405
	Silurian		436
	Ordovician		510
	Cambrian		560
Pre-Cambrian			4,600

Figure 1.5 Grapevines growing on steep slopes along the shore of Lake Geneva in the central part (Lavaux) of the Vaud region, Switzerland.

and on the upper slopes of the Rhine River in the Rheingau region, Germany, a premier region for Riesling wines. The soils on such slopes are very shallow and well drained unless they contain glacial till or in some cases loess (windblown material).

Transport of weathered materials by water or ice is a most important means of distributing parent material for subsequent soil formation. At times during the Pleistocene epoch (see box 1.2), much of the present-day wine regions of Europe was covered by glacial ice that moved weathered rock materials—fragments and finely ground "rock flour"—over the landscape. Relatively young soils (<10,000 years of age) have formed on these deposits of glacial till but are usually not well suited to vineyards because of poor drainage. During the interglacial (warm) periods of the Pleistocene, however, glaciers retreated and, as they melted, large amounts of melt water containing a heterogeneous mix of rock material swept down rivers and formed glaciofluvial deposits. Some of the best vineyard regions have developed on the better drained soils formed on such deposits, as in the Médoc region and southern Rhone Valley in France and the Bio-Bio, Maule, and Rapel Valleys in Chile (figure 1.6).

Over most of the land surface, erosion of weathered material by water has produced areas of alluvial deposits in the form of extensive plains and river

Figure 1.6 Soil formed on a glacial outwash deposit overlying older weathering granite in the Cauquenas region, Chile. The scale is 15 cm. (White, 2003)

terraces. The finer sedimentary material is carried farthest from the river before settling out, whereas the coarser sand particles (up to 2 mm diameter), gravel (2 to 60 mm), stones and boulders (>60 mm) are deposited progressively nearer the river, usually forming terraces. Soils formed on better drained alluvial deposits are used extensively for vineyards in the Riverina region and Riverland-Sunraysia region, Australia. Examples of gravelly alluvial deposits under vineyards occur in of New Zealand's Hawke's Bay region, as on the Gimblett Gravels, well known for Cabernet Sauvignon and Syrah (Shiraz) wines, and the Marlborough region, renowned for its Sauvignon Blanc (figure 1.7).

Soils on Sandstone

Sandstone, as the name implies, is made up of grains of sand comprising minerals such as quartz, feldspar, or simply rock fragments in the size range 0.06 to 2 mm. Classed as sedimentary rocks, formed by the consolidation of original deposits of weathered materials moved by wind or water, sandstones generally form soils of low fertility because of a paucity of weatherable minerals. When sandstone has been weathering for a long time under rainfall >1,000 mm, deep

Figure 1.7 A shallow soil formed on a gravel and boulder alluvium in the Marlborough region, New Zealand. The scale is 10 cm.

soil profiles of uniform texture similar to that shown in figure 1.8 can form. Under native vegetation, complexes of iron (Fe) and aluminum (Al) with organic compounds released from decomposing plant litter are gradually leached into the subsoil (B horizon) where they are transformed into Fe oxides (orange-red colors) and humified organic matter (black color). The A horizon above the subsoil is conspicuously bleached due to the removal of Fe and organic matter. Although such soils are extremely well drained because of their coarse texture, they are not much used for viticulture because of their poor water holding capacity and very low fertility.

Where the sedimentary rocks have a greater proportion of particles smaller than sand size—the siltstones and mudstones—duplex soil profiles similar to that shown in figure B1.1.1 can form. Many vineyards in southeast Australia, particularly in the cooler regions, have been established on such soils, which also occur in the Stellenbosch region of South Africa on granitic and sedimentary parent materials. A characteristic of many duplex soils, especially under a cool and wet winter climate, is gleying and mottling in the subsoil where, because of the high clay content, drainage is impeded. The biochemical processes of gleying and mottling are explained in box 1.3. Although grapevines are naturally deep rooting,

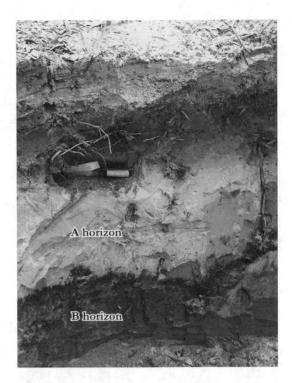

Figure 1.8 A deep Podzol soil formed on sandstone near Sydney, Australia. Iron complexed with organic matter has been translocated from the bleached horizon, to be deposited as orange-red iron oxide in the subsoil. The pale soil on top is spoil from the pit. See color insert.

inhospitable subsoils and poor drainage can inhibit this rooting. However, the constraint of poor drainage can be managed by deep ripping with or without applications of gypsum, as discussed in "Preparing the Site by Cultivation and Ripping," chapter 2.

Soils on Igneous Rocks

Igneous rocks, produced from volcanic activity deep in the earth's crust (intrusive or plutonic rocks) or at the earth's surface (extrusive rocks formed from lava flows and ash showers), are the ultimate source of all rocks on the planet (see table 1.1). They range from silica-rich, "acidic" rocks such as granite and rhyolyte to rocks of intermediate basicity (dacite and andesite) to low-silica "basic" rocks such as basalt, gabbro, and dolerite. This range of rock types naturally gives rise to a variety of soils. For example, in the Granite Belt region of southeast Queensland, Australia, the landscape is old and much reworked by erosion and deposition: hence the soils formed are acidic (pH(H_2O) <6) and of variable depth but consistently coarse-textured, reflecting the high quartz content of the granitic parent material (figure 1.9). Such soils are generally well drained.

Box 1.3 The Processes of Gleying and Mottling

The weathering of Fe-containing minerals releases ferric (Fe^{3+}) and ferrous (Fe^{2+}) ions, which are in a reversible equilibrium (a redox reaction) that depends on the availability of oxygen (O_2). High O_2 concentrations (characteristically aerobic conditions) favor Fe^{3+} compounds, particularly the oxides (general formula Fe_2O_3.nH_2O, where "n" is a variable number), which are insoluble and orange-red ("rusty") in color. Low O_2 concentrations (anaerobic conditions) favor Fe^{2+} compounds, which are more soluble than Fe^{3+} compounds and give a blue-gray color to the soil. Soils that are waterlogged most of the time are predominantly blue-gray in color and have little or no structure. This condition is described as "gleyed" and typically occurs in clay subsoils that remain saturated for most of the year. Figure B1.3.1 shows an example of a vineyard where the vines have died from soil waterlogging.

Figure B1.3.1 A vineyard soil with an impermeable subsoil in Virginia. The vines died from waterlogging. (Photo courtesy of Brad Johnston, Warrenton, Virginia.)

However, in soils where waterlogging is periodic (e.g., occurring mainly during wet periods in winter) the soil matrix is not uniformly gleyed but instead shows "mottling." Mottling occurs when dissolved Fe^{2+} ions migrate to pockets of higher O_2 concentration where they are oxidized to Fe^{3+} and precipitate as Fe oxide. The rust-colored Fe oxide mottles show up against the gley-colored background,

(continued)

Box 1.3 *(continued)*

as shown in the exposed subsoil of a duplex soil in the Sunbury region, Victoria, Australia (figure B1.3.2). Note how the clay of the subsoil has been smeared by the action of the digging machine. The extent of gleying and mottling enables a visual appraisal to be made of the quality of drainage in a soil profile.

Figure B1.3.2 This orange-mottled soil occurred in the subsoil of a duplex soil in the Sunbury region, Victoria, Australia. Note smearing of the moist clay by the excavator. See color insert.

On the other hand, soils formed on more basic rocks are finer textured (higher clay content), of neutral to alkaline pH, and generally well structured due to their aggregates being stabilized by Fe and Al oxides (Fe_2O_3 and Al_2O_3). Figure 1.10 shows a gradational red soil formed on volcanic parent material of the Cambrian period in the Goulburn Valley region, Victoria, Australia, under an annual rainfall of 554 mm. This parent material also outcrops under a similar rainfall in the Heathcote region, some 100 km to the southwest, which has a reputation for producing intensely flavored Shiraz wines. Under a cooler, wetter climate, basic rocks of basalt and dolerite have weathered to form deep red loams, as in the Willamette Valley, Oregon, parts of the Yarra and King Valley regions in Victoria, Australia, and in northeast Tasmania (figure 1.11). The Willamette Valley region is noted for its Pinot Noir wines. Although the combination of high rainfall, natural soil fertility, and deep rooting often predisposes

Figure 1.9 A deep coarse-textured soil formed on granitic parent material.

to a problem of excess vigor in vines planted on such sites, this can be controlled by canopy management and the correct choice of grapevine rootstock (see "Rootstocks," chapter 5).

This brief overview of the effect on soil formation of parent material, interacting with the other factors of climate, organisms, relief, and time, gives some indication of the enormous variability that can be shown by soil in the landscape. The extent of this variability and its influence on viticulture over the centuries are discussed in the following section.

Soil Variability and the Concept of *Terroir*

Some of the earliest examples of winemakers acknowledging the interaction of soil variability with vine performance and wine character go back to the Middle Ages (from the 8th century AD) in the Burgundy and Rheingau regions, where monasteries possessed large tracts of land on which vines were grown and wine made. Particularly notable were monasteries at Cluny in the Mâconnais region, France, Clos de Vougeot at Cîteaux in the Côte d'Or, Burgundy region, and Kloster Eberbach in the Rheingau region (Johnson and Robinson, 2013),

Figure 1.10 The gradational profile of a Brown Earth under vines in the Goulburn Valley region, Victoria, Australia. There is a band of precipitated calcium carbonate at depth. See color insert.

where successive generations of monks identified, by trial and error, the special character of wine made from individual parcels of land (*goût de terroir*). Cultivars were winnowed out until the first choice was Pinot Noir in the Côte d'Or and Riesling in the Rheingau. Gamay was found unsuitable on the soils of the Côte d'Or and "banished" by order of Duke Philippe of Burgundy in the late 14th century but subsequently found its natural home on the schist and granite soils of the Beaujolais region to the south. Subsequently, Chardonnay was planted on the hard limestone soils of the southern Côte d'Or, extending into the Chalonnais region, and farther south to parts of the Mâconnais region where the beautifully structured Brown Earths around the Roche de Solutré and Roche de Vergisson produce excellent white wines (figure 1.12).

Having identified the best combinations of cultivar and soil, the winemakers strove to maintain consistency in quality and character for individual labels by blending wine from small blocks in different proportions, depending on the vintage. Thus were laid the foundations of the concept of *terroir*, which expresses in one word the complex interaction of soil, local climate, cultivar, and the winemaker's skill in determining the character or individual "personality" of a wine. After the French Revolution of 1789, Napoleon ordered the large monastic

Figure 1.11 A deep red loam on colluvial basalt in a vineyard in the Willamette Valley region, Oregon. See color insert.

landholdings to be broken up, with the result that vineyards were fragmented into 0.5 to 1 ha blocks (called *climats*) and acquired by many owners. However, much of the unofficial classification of wine from the *climats* in Burgundy was preserved and incorporated in 1861 into an official Appellation d'Origine Contrôllée (AOC) system, which still exists today. Now there are several hundred appellations in the greater Burgundy region.

The French AOC system expresses the concept of *terroir* through an identification of viticultural areas (regions), based on the distinctive character of wines traditionally produced in those regions. The AOC system prescribes the cultivars, viticultural methods, maximum yield, and wine alcohol content that are acceptable in a particular appellation, and the wines produced are allocated to four categories from the top down—Appellation Contrôllée, Vin Délimité de Qualité Supérieure, Vin de Pays, and Vin de Table.

Despite the reputation of many of the wines in top categories of the AOC system, there is much wine of ordinary quality marketed in the lower categories. An imbalance of supply and demand and fierce international competition have led to a relaxation of AOC regulations in some French wine regions and in similar regulatory systems in force in Italy and Spain. Also, New World countries have

Figure 1.12 A well-structured Brown Earth near the Roche de Solutré in the southern Mâconnais region, France. The scale is 10 cm.

adopted systems of wine-region classification, such as the American Viticultural Areas, the Australian Geographical Indications, and the Wine of Origin Scheme in South Africa, which to varying degrees are built on the underlying principles of *terroir*, or "sense of place."

The concept of a site's *terroir* being expressed through the distinctiveness of wines produced from that site has been reinforced in recent times through the organic/biodynamic viticulture movement (discussed in chapter 3). Especially in biodynamic viticulture, the proponents argue that only by adopting "natural" practices of soil management and minimum intervention in the vineyard can the uniqueness of a site be expressed—"have the confidence to listen and allow the site to speak to you" is the advice of Ted Lemon of Littorai Wines, Sonoma County, California (Halliday, 2013). On the other hand, various scientific studies have failed to identify specific causal relationships between one or more soil or geological properties and a site's *terroir*, as argued at a recent Geological Society of America conference in Portland, Oregon (Maltman, 2009). A possible exception to this generalization is the relationship between the rate of soil water supply, the performance of dry-grown vines, and wine character that has been extensively studied in the Médoc and St. Emilion regions of the Bordeaux region. Despite its lack of scientific underpinning, the contribution of soil to a site's *terroir* and wine "typicity" is of great interest to many winemakers who are seeking a competitive advantage in fiercely competitive national and international wine markets. We return to this topic in chapter 6. However, in the absence of conclusive scientific knowledge, a pragmatic approach is to think of *terroir* as embodying an empirical but imperfect classification of soils specifically for viticulture, which has been honed by the experience of practicing vignerons over a long time.

Soil Classification and *Terroir*

Classification involves the "packaging" of soil variability so that soils with similar values of a property such as profile form, color, and texture are grouped together and are separated from groups of soils with dissimilar property values. To assess the full range of soil variation in an area of interest, a survey is conducted. The variation is divided into "classes," which collectively comprise a classification. Soil scientists have justified traditional classification of this kind by saying it is valuable for communicating soil information nationally and internationally. However, experience shows that information in this form is not widely used by farmers, winegrowers, and land managers, in part because detailed information about a particular location is lost in the creation of the general-purpose classes and in part because scientists have produced several classifications that have resulted in a confusing suite of soil names. In Australia, for example, four national soil classifications have been used at various times over the past 60+ years, culminating in the Australian Soil Classification (Isbell, 1996). Following its revision (Isbell, 2002), the system has been more widely accepted as a truly national classification and its soil orders, representing the highest categorical level, correlated with putative international classifications, such as Soil Taxonomy (Soil Survey Staff, 1999). However, the older Great Soil Group Classification still retains currency in the viticulture industry, and its Great Group names are used when necessary in this book.

Recognition of *terroir* in Old World vineyards, especially in Burgundy and other historically important regions of France, relies on an accumulated knowledge of soil properties relevant to vine performance and fruit quality on a scale of meters rather than kilometers. In Australia, where the wine industry is much younger than in France, the need to provide relevant soil data for grapevine culture led to the creation of a special-purpose classification, which as far as possible uses nontechnical terms to describe soil properties observable in the vineyard (Maschmedt et al., 2002). Soil survey is the process by which such information is gathered.

Soil Survey and Mapping

The Traditional Method

The traditional method of soil survey is normally used for general-purpose soil assessment and is carried out at varying intensities, ranging from reconnaissance (scale 1:2,000,000 to 1:100,000) to detailed (1:50,000 to 1:1000).[3] In this case, soil survey may start with an examination of air photographs to identify possible soil changes from landscape features, for example, stream courses, slopes and valleys, and changes in land use and vegetation. Notional boundaries between soil classes are identified and checked by free survey on the ground when samples are taken with an auger or soil corer. Figure 1.13(A) shows examples of manually operated soil augers and corers; a more versatile hydraulic soil corer that can drill down to 1.3 m is shown in figure 1.13(B). Once the boundaries separating classes have been confirmed, rectangular pits are dug to expose the soil profile thought to be representative of each class. Figure B1.1.1 is an example of a soil profile pit. The pit allows the visual characteristics of each horizon to be described accurately; these include sharpness of boundaries, depth, color and mottling, presence of organic matter, stones, structure, and presence of carbonate or other salt deposits, as well as the "feel" or texture of the soil (see "Texture Analysis and Calibration for Texture," chapter 2). Samples are also collected from each horizon for laboratory analysis.

The output of such a survey is a map that shows the spatial distribution of soil classes. The degree of homogeneity of each class depends on the intensity of soil sampling and the scale at which the class distribution is mapped. If the soil distribution is relatively simple or the map scale is large (>1:5000), individual classes can be displayed. However, if the distribution is complex (as it often is with alluvium-derived soils) or the map scale small, two or three classes may be grouped into a mapping unit. From such a conventional survey, which is usually linked to a general-purpose soil classification, the key properties of a soil typical of each class are identified. This typical soil is called a soil type, and the properties of each type are supplied in a legend that accompanies the soil map. Figure 1.14 shows an example of a typical soil map, derived from a general-purpose soil survey, which shows the dominant soil classes in plain colors as well as the complexes of codominant and subdominant classes in hashed colors. Because such maps are rarely available at a scale larger than 1:5000 (and more commonly at 1:50,000 or 1:100,000), much detail about soil variation over distances of several meters is lost.

In Australia, when a soil survey is required for a new vineyard or the redevelopment of an existing vineyard, a standard practice has been to locate pits on a 75 × 75 m grid, without necessarily any recourse to air photographs, and for the

---

[3] A scale of 1:1000 means that 1 cm on the map represents 10 m on the ground.

Figure 1.13 (A) Examples of augers and corers for soil sampling. (B) A tractor-mounted, hydraulic soil corer capable of sampling down to 1.3 m. (Photo courtesy of Martin Peters, Farming IT, Australia.)

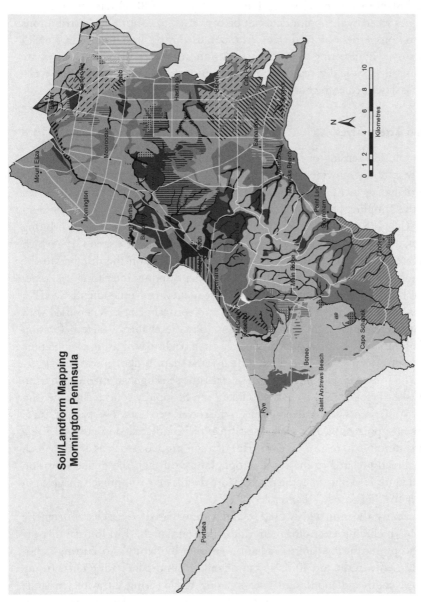

Figure 1.14 Soil map of the Mornington Peninsula region, Victoria, Australia, based on the general-purpose Australian Soil Classification. The map scale is 1:100,000. (Map from the Victorian Resources Online website, hosted by the Victorian Department of Environment and Primary Industries.) See color insert.

exposed soil profiles to be described. The data recorded are primarily soil depth, color, stoniness, field texture, and the presence or absence of $CaCO_3$. The results are usually displayed not as a map but as point profile diagrams that are not necessarily linked to any soil classification. Bramley et al. (2010) refer to statistical techniques whereby this point data can be converted to continuous distributions of the soil properties observed. However, because of the limitations of a 75 ×75 m grid survey in detecting short-range variation in a soil property, such maps are not as useful as those derived from high-resolution sensing and calibration techniques, as described in the next section.

Modern Alternatives

High-resolution sensing of soil properties, either at close range (proximally) or remotely, can now provide soil information at an appropriate scale for intensive vineyard management. Soil properties are "sensed" in a dense spatial array in which the sampling points are accurately located by a global positioning system (GPS). The GPS relies on fixing a point on the surface with reference to orbiting satellites, a system in common use being the differentially corrected GPS, which is accurate to less than ±50 cm in Australia, depending on the distance to the base station. However, a more accurate instrument finding increasing use, especially in broad-acre agriculture, is the dual-frequency, real-time kinematic GPS, which is accurate to 2 cm in the horizontal and vertical planes. When used with a height reference level, the real-time kinematic GPS enables a digital elevation model of the site to be constructed. This can then be displayed using a geographic information system (GIS) in a three-dimensional map. When an real-time kinematic GPS is used in combination with a real-time sensing instrument such as an EM38 (see the following), vineyard soil data can be rapidly collected at a density of 200 to 600 points per ha, depending on the travel speed and row spacing. These data are interpolated using a special statistical technique called kriging, to give a high-resolution map (e.g., a 2 × 2 m grid) of a soil property, such as depth, which can be directly related to the surface topography through the digital elevation model. Figure 1.15 illustrates the result. More details of this approach are given in "Starting the Soil Survey," chapter 2.

Currently, the most widely used soil sensors are based on the electromagnetic spectrum, including electromagnetic induction (as in the EM38), gamma-ray spectroscopy measuring gamma radiation emitted by naturally occurring radioactive elements in the top 30 to 45 cm of soil, ground-penetrating radar (using long-wavelength radiation), and laser-imaging radar (using ultraviolet, visible, and near-infrared radiation). Electronic signals from the sensors are streamed to a logger, where the data are stored in digital format for subsequent downloading to a computer and processing in a GIS. A GIS is specialized computer software in

Soil depth (mm)
- < 300
- 300 - 350
- 350 - 400
- 400 - 450
- 450 - 500
- 500 - 550
- 550 - 600
- > 600

Figure 1.15 Three-dimensional map showing surface topography and soil depth in a 7.3 ha vineyard in the Coonawarra region, South Australia. (Map courtesy of Dr. Robert Bramley, CSIRO Ecosystems Sciences, Adelaide, South Australia.)

which spatially referenced soil data are stored in virtual layers, along with information on field boundaries, streams, and roads, and are linked to attribute data that describe each spatial feature in the GIS.

The raw sensor output can be manipulated to produce a map showing the spatial variability in soil in the area surveyed, without knowing the cause of the variability. With this information, the surveyor can decide on the best placement of inspection pits to encompass the full range of soil variability. However, in order to understand the meaning of this variation in terms of a soil property of interest, the sensor output must be converted to actual values of the property by calibration against laboratory-measured values, as in the case of clay content or salinity (see "Texture Analysis and Calibration for Texture," chapter 2), or by using a model of soil chemical reactions, as in the case of estimating a soil's lime requirement (see box 3.6, chapter 3). Once a calibration relationship is established, spatially referenced data can be collected over adjacent areas, stored in a GIS, and used directly without the need for filtering through a soil classification. Provided proper calibrations are established, data for soil depth, soil texture, permeability, salinity, and so on can be used to plan the layout of a new vineyard, including the allocation of individual varieties to particular sites and the design of an irrigation system.

Even in established vineyards, more precise knowledge of soil variation enables pruning, irrigation, fertilizing, and mulching to be carried out on a zonal basis, where a zone is an area in which the soil is less variable than in the vineyard as a whole. Note that the boundaries between zones in an established vineyard may or may not coincide with those between vineyard blocks. Because soil variation can be correlated with vine vigor, fruit yield, and quality, it can be to the winemaker's advantage for these zones to be harvested separately, allowing the making of wines of different styles or for different price points (see "Managing Natural Soil Variability in a Vineyard," chapter 6). Thus the application of modern technology potentially closes the circle that began with the identification of particular *terroirs* in monastic vineyards on the slopes of the Côte d'Or centuries ago.

Regardless of whether or not vignerons pursue a *terroir* approach to winegrowing, knowledge of soil properties, their variability, and how the soil can be managed to achieve production objectives is an essential component of vineyard management. Chapters 2 to 5 identify and describe the important soil properties and soil management practices for viticulture. Chapter 6 reviews this soil knowledge and considers how it can be used to achieve particular production objectives.

Summary Points

1. In popular parlance, the term "soil health" has replaced "soil quality," which refers to a soil's fitness for purpose. Soil health or soil quality is an integral expression of a soil's chemical, physical, and biological attributes, which determine how well the soil supports productive plant growth and provides a variety of ecosystem services.
2. Broadly, both inherent and dynamic factors contribute to soil health. The inherent factors are primarily the geology (parent material of the soil), climate, organisms living on and in the soil, and topography, which do not change significantly on a human time scale, and the age of a soil. On the other hand, dynamic factors such as soil organic matter, pH, nutrient availability, soil structure, and water supply can change much more quickly and be manipulated by human intervention.
3. Because of the complex interaction of inherent and dynamic factors over time, the distribution of soils in the landscape is highly variable. In Europe, most notably France, the varying environmental endowments of a site, in combination with human influences acting through the selection of grape varieties and viticultural practices, are said to constitute the site's *terroir* or "sense of place," conferring a distinctive character on the wine produced.
4. As parent material weathers in a variable environment, soil profiles develop with distinctive visual and behavioral characteristics that give rise to a range of soil types. A comparison of profiles of vineyard soils formed on different rock types such as schist, limestone, sandstone, granite, basalt, and transported materials derived from these rocks illustrates this variability.

5. Soil variability is measured by real-time sensing of soil properties remotely or close at hand (proximally) at a high spatial resolution. The observations can be calibrated by measurements made by soil surveyors examining soil in profile pits or from augered samples (laboratory analysis). Storage of the data in a GIS and its visualization allow more precise maps of the spatial structure of soil variability to be produced than is possible with traditional maps based on a general-purpose classification.

6. Armed with this knowledge, winegrowers can make better decisions about vineyard management, in particular how to grow vines to meet their objectives of achieving a specific fruit yield, quality, wine style, and market price point.

2

Site Selection and Soil Preparation

Determining the Site

As outlined in chapter 1, "determining the site" in old established wine regions such as Burgundy, Tuscany, and the Rheingau has been achieved through centuries of acquired knowledge of the interaction between climate, soil, and grape variety. Commonly, vines were planted on the shallow soils of steep slopes, leaving the more productive lower terraces and flood plains for the cultivation of cereal crops and other food staples, as shown, for example, by the vineyards along the Rhine River in Germany (figure 2.1). The small vineyard blocks of the Rhine River, the Côte d'Or, Valais and Vaud regions of Switzerland allowed winegrowers to differentiate sites on the basis of the most favorable combination of local climate and soil, which underpinned the concept of *terroir*. In much of the New World, by contrast, where agricultural land was abundant and population pressure less, vineyards have been established on the better soils of the plains and river valleys, as exemplified by such regions as the Central Valley of California, the Riverina in New South Wales, Australia, and Marlborough in New Zealand. Apart from the availability of land, the overriding factor governing site selection was climate and the suitability of particular varieties to the prevailing regional climate. In such regions, although soil variability undoubtedly occurred, plantings of a single variety were made on large areas and vineyard blocks managed as one unit (figure 2.2). Soil type and soil variability were largely ignored.

Figure 2.1 Vineyards on steep slopes above the Rhine River, Germany.

Figure 2.2 Extensive vineyards under uniform management in the Riverina region, New South Wales, Australia.

Notwithstanding this approach to viticulture in New World countries, in recent time winegrowers aiming at the premium end of the market have become more focused on matching grape varieties to soil and climate and adopting winemaking techniques to attain specific outcomes for their products. For established vineyards, one obvious result of this change is the appearance of "single vineyard" wines that are promoted as expressing the sense of place or *terroir*. Another reflection of this attitudinal change is the application of precision viticulture (see "Managing Natural Soil Variability in a Vineyard," chapter 6), whereby vineyard management and harvesting are tailored to the variable expression of soil and local climate in the yield and sensory characteristics of the fruit and wine. For new vineyards, however, determining the site starts with gathering as much data as possible on

- Climate
- Soil types and their distribution
- Availability of water for irrigation
- Potential for pest and disease attack

Together, these factors interact to determine a site's suitability for different grape varieties and the potential profitability of the winegrower's business.

Climate

Three levels of climatic influence are recognized in viticulture:

1. The macroclimate or regional climate: largely determined by latitude and altitude and applicable over tens of kilometers but modified by the distance from moderating influences such as the sea or large lakes.
2. The mesoclimate or site climate: more local than the macroclimate and primarily determined by sharp changes in altitude over short distances, as well as by slope and aspect. An example is the difference in mesoclimate between north- and south-facing slopes along the Rhine River in Germany. In California, the macroclimate of the Lodi District in the Central Valley is different from that of the Napa region, but within each region there are several mesoclimates, depending on the distance from San Francisco Bay or closeness to the Mayacamus and Vaca Mountains.
3. The microclimate or canopy climate: the climate within and immediately around the vine canopy.

Various bioclimatic indices have been developed to classify the macroclimate and help to identify the best regions for particular "cultivars".[1] These indices are

---

[1] Cultivar is a shorthand term for cultivated varieties.

briefly described in box 2.1. As an example of the use of these indices, figure 2.3 shows a diagram of the ranges in growing season temperature that are considered optimum for the production of premium quality wines from some of the best known varieties. Such a diagram needs to be used with caution, however, because ongoing climate change may lead to a number of possible outcomes, such as

---

Box 2.1 Climate Indices for Site Selection

One of the earliest bioclimatic indices for wine grapes, developed in California by Amerine and Winkler (1944), is based on the summation of heat units during the growing season, that is, for seven months from October to April in the southern hemisphere or April to October, inclusive, in the northern hemisphere. The summation, called heat degree days (HDD), for a site is calculated as follows:

HDD = Sum of [number of days × (mean daily temperature − 10°C)]

The mean daily temperature is the average of the daily maximum and minimum temperature, which should be derived from records for at least 10 years. The 10°C (50°F) cutoff was chosen because few grape cultivars grow actively below this temperature, which is why the term growing degree days, synonymous with HDD, is used in some countries. Five macroclimatic regions are recognized, ranging from Very Cool (<1390 HDD, temperature in °C) to Very Warm/Hot (>2220 HDD, temperature in °C).

Another system developed in Australia by Smart and Dry (1980) is based on the correlation between mean January temperature and HDD for places of similar continentality. The continentality index is measured by the difference between the mean January and mean July temperatures (for the southern hemisphere). More recently, Smart (2001) has promoted the homoclime concept, which is a way of comparing statistically the mean monthly temperatures and rainfall for a site with those for established wine regions around the world. For sites without a record of climate data, the climate can be estimated from national meteorological databases providing data on a 2 × 2 km grid. After the homoclime of a site has been established, the most suitable cultivars can be recommended, based on experience in other parts of the world (see www.smartvit.com.au). Figure 2.3 illustrates the application of Jones's (2006) growing season temperature, global climate classification for premium wine production.

Because the HDD concept is relatively crude, Gladstones (1992, 2011) developed a more refined temperature index, expressed as a summation of "biologically effective" degree days (E° days). Based on the strong dependence of vine phenology (charting a vine's progression from bud burst to maturity), Gladstones estimated E° from the same starting point as HDD but with adjustments for the diurnal temperature range, day length (dependent on latitude), and an upper temperature limit of 19°C. Full details of this more complex index are given in chapter 11 of Gladstones (2011).

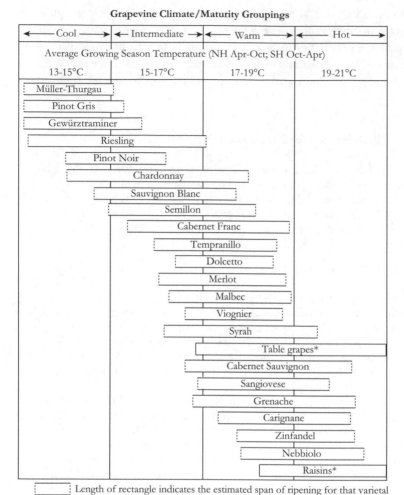

Grapevine Climate/Maturity Groupings

◄──── Cool ────► ◄── Intermediate ──► ◄──── Warm ────► ◄──── Hot────►

Average Growing Season Temperature (NH Apr-Oct; SH Oct-Apr)

13-15°C   15-17°C   17-19°C   19-21°C

Müller-Thurgau
Pinot Gris
Gewürztraminer
Riesling
Pinot Noir
Chardonnay
Sauvignon Blanc
Semillon
Cabernet Franc
Tempranillo
Dolcetto
Merlot
Malbec
Viognier
Syrah
Table grapes*
Cabernet Sauvignon
Sangiovese
Grenache
Carignane
Zinfandel
Nebbiolo
Raisins*

▢ Length of rectangle indicates the estimated span of ripening for that varietal

Figure 2.3 Climate-maturity groupings based on growing season average temperatures. The horizontal bars represent the range of temperatures for which each variety is known to ripen and produce high to premium quality wine in the world's benchmark regions. The dashed line at the end of the bars indicates that some adjustments may occur as more data become available, but changes of more than +/- 0.2°–0.5°C are highly unlikely (Jones et al., 2012).

- an increase in mean growing season temperature that varies from region to region globally
- a more variable macroclimate and mesoclimate, with an increased frequency of unusually hot days and unseasonal frosts
- changes in the amount of rainfall, its distribution between regions, and its seasonal incidence

Any one of these could affect the suitability of a given cultivar for a particular region or subregion.

Temperature records around the globe show that the average surface temperature has increased by about 0.8°C during the 20th century, with most of this increase occurring in the last 50 years of that century. Climate scientists have used global circulation models (GCMs) to simulate this warming trend, which has been attributed to an increase in the atmospheric concentrations of greenhouse gases, produced mainly by human activities. Using these models, projections have been made of future temperature increases and of other changes, as suggested in the aforementioned points. Some winegrowing companies in Australia and overseas have acted on these projections and bought vineyards in cooler maritime regions, or at higher elevations, in an attempt to mitigate any adverse effects of climate change.

Nevertheless, it must be remembered that forecasts, for example, of an average temperature increase of 0.4° to 2°C[2] by 2030 are simply model projections, with inherent uncertainties depending on how well the climate system, with its complex biophysical processes and feedback mechanisms, is understood and what the actual values of the numerous input variables will be during the next 17 years. Indeed, a recent analysis of the transient climate response to forcing factors between 1970 and 2009 shows the actual temperature increase to be tracking at the lower level of the ensemble of GCM projections (Otto et al., 2013). Similarly, Hawkins (2013) concluded from a comparison of 42 GCM projections with global temperature records that "global temperatures have not warmed as much as the mean of the model projections in the past decade or so," an observation that is consistent with the "flat" trend in the earth's surface temperature for the past 15 years (Anon., 2013). Climate change and its potential impacts are discussed in more detail in "Climate Change and Possible Consequences" in chapter 6.

The effect of climate change on vine performance is not only confined to an effect of temperature increase during the growing season on vine growth and fruit ripening but also can involve the effect of higher atmospheric CO_2 concentrations on vine growth, possible increased stress in water-limited regions, and possible changes in pest and disease pressure. The interaction of these climate-related uncertainties reinforces the concept that site selection should be based on a more comprehensive suite of factors than climate alone, as discussed in the next section.

---

[2] The wide range in the projected increase is due to the different GCMs used to make the projections.

Site Selection Indices

A more complete approach to site selection than that of climate analysis alone involves the development of a local site index (LSI), which combines knowledge of the requirements of particular cultivars for the factors listed under "Determining the Site." Viticultural and soil experts are asked to identify the key biophysical properties and weight them to form a composite index for sites in a region. For example, when applied to assessing site suitability for wine grapes in the West Gippsland region, Victoria, Australia, soil, climate, and topography were weighted in the ratio 70:20:10 (reflecting the perceived order of their importance; Itami et al., 2000).

A similar approach was used in the hilly and mountainous terrain of central Virginia,—a region colder than West Gippsland. There, frost in spring and freezing in winter were serious constraints, so the most important factor in the LSI was altitude (affecting temperature). Altitude was weighted at 30%; followed by soil, 25%; existing land use, 20%; slope, 15%; and aspect, 10% (Boyer and Wolf, 2000). Another example of an LSI is that developed for Cabernet Sauvignon in the Hawke's Bay region, New Zealand, which has a maritime climate of cool winters and warm to hot summers (Tesic et al., 2002). In this index, soil properties such as maximum rooting depth, topsoil gravel content, and soil texture (affecting water-holding capacity and soil temperature) were combined with climatic variables, such as growing season rainfall and mean October and January temperatures. The resultant index was better correlated with growth characteristics and fruit quality than several purely climatic indices.

The LSI approach has been used to develop the concept of digital *terroirs*, which are described in chapter 6. A grower is more appreciative of such indices if the primary data are spatially defined using a global positioning system (GPS), entered into a geographic information system, and displayed in a colored map.

Although "determining the site" is only the first step in a chain of events in wine production, it is important because suitable land is becoming scarce in many of the more desirable wine regions. Also, the investment per hectare (ha) in establishing a new vineyard is large, so mistakes made in site selection can ultimately be costly. The remainder of this chapter focuses on the soil factor in site selection; it presents methods of gathering soil data and modifying the soil–vine environment to optimize vineyard performance.

Steps in Site Selection

Topography overall is important in site selection in respect of cold air drainage, susceptibility to frost, and row orientation for safe use of machinery, as discussed later. However, as indicated in "Site Selection Indices," the soil also plays an important part in determining a site's suitability for a vineyard. The most direct effect is on vine rooting and whether there is any serious chemical, physical, or biological factor that will impede root development and affect the health of vines and their ability to withstand adverse conditions. The winegrower will also want to know whether the nature of the site and soil may predispose to undesirable off-site effects through erosion and/or leaching of nutrients and pest-control chemicals. Consider an apparently uniform area of flat land to be planted to vines, such as is shown in figure 2.4. The question arises whether the soil below is a uniform, well-drained red loam, as shown in figure 2.5A, or whether the subsoil is poorly drained, as shown in figure 2.5B, or whether there is some other factor such as excessive acidity or salinity that could inhibit growth. Only a soil survey can provide answers to these questions.

Starting the Soil Survey

A soil survey allows soil limitations to vine growth to be identified and located, plans to correct any problems to be made, areas for particular varieties to be decided, and the layout of an irrigation system, if required, to be designed. The section on "Soil Survey and Mapping" in chapter 1 introduces some traditional and modern methods of soil survey for vineyards.

Figure 2.4 An apparently uniform area of land newly prepared for planting vines in the Riverina region, Australia.

(a)

(b)

"Mottled" zone

Figure 2.5 (A) Deep red loam soil in the Heathcote region, Victoria, Australia. The scale is 15 cm. (B) A vineyard soil in the Hawkes Bay region, New Zealand, with an impermeable subsoil that experiences intermittent waterlogging. See color insert.

Box 2.2 Electromagnetic Techniques for Soil Survey in Vineyards

The EM38 measures the bulk soil *EC*, which is affected by salinity, water content, clay content, bulk density, and soil temperature. However, if the in-field variation in one of these factors is large compared with the others, the EM38 can be calibrated to measure that dominant factor. A basic instrument is shown in figure B2.2.1A. It is normally placed in an insulated case to minimize temperature

(a)

(b)

Figure B2.2.1 (A) An EM38 in vertical orientation (White 2003). (B) An EM38 on a nonconducting sled being pulled through a vineyard. (Courtesy of Dr. Robert Bramley, CSIRO Ecosystem Sciences, Adelaide, South Australia.)

(continued)

Box 2.2 *(continued)*

variations and mounted in front of a tractor or placed on a rubber or Perspex sled drawn by an all-terrain vehicle (figure B2.2.1B). A GPS mounted on the vehicle enables the position of the sensor to be tracked as it is moved over the land.

The EM38 works by EM induction, whereby a low-frequency, oscillating electric current in a transmitter generates small, fluctuating currents, with associated magnetic fields, in the soil. The ratio of secondary to primary magnetic field strengths is directly proportional to the *EC* of a roughly cylindrical soil volume below. These magnetic fields are detected by receivers built into the instrument. The current "standard" EM38–MK2 has two receiver coils, separated by 1 m and 0.5 m from the transmitter, providing *EC* data from depth ranges of 1.5 m and 0.75 m, respectively, when positioned in a vertical orientation, and 0.75 m and 0.375 m, respectively, when in horizontal orientation (see www.geonics.com and www.farmingit.com.au).

A related technique is that of measuring soil electrical resistivity (the reciprocal of conductivity). A direct electric current is applied to electrodes built into the coulter wheels of a cultivator drawn behind an all-terrain vehicle. The potential difference between the electrodes is related to the soil's resistivity.

In Australia, the EM38 is used to measure vineyard soil variation. The results have been calibrated against salinity, clay content (also reflecting water content), and soil depth (e.g., where a sharp change from soil to rock occurs within the depth range of the instrument). When the EM or a resistivity instrument is used in an existing vineyard, care must be taken that the signal is not distorted by steel posts and wires. This is not a problem for the EM38 when the row spacing is more than 2.5 m, provided the instrument follows the midline between rows, and it is less of a problem for electrical resistivity measurements because the electrodes are closer together than the transmitter and receiver in an EM38.

Commonly, a series of preliminary holes is "augered" to determine whether a full survey with sensing devices and profile pits is needed. If a full survey is deemed necessary, the first step may involve remote sensing, such as with air photos or satellite images, or with techniques such as gamma radiation and laser-imaging radar. Traditionally, stereoscopic analysis of air photos (giving a three-dimensional picture) has been used to map surface topography and vegetation, both of which can indicate soil changes, but nowadays a real-time kinematic GPS is used to construct a digital elevation model of the site (see figure 1.15, chapter 1).

The next step usually involves proximal sensing of the soil with an EM38, as described in box 2.2. At its most basic, an electromagnetic (EM)-derived map shows the pattern of soil variation that can be caused by a number of factors. If several soil properties (e.g., soil depth, clay content, and salinity) are involved in determining the EM signal, the main value of the map is in indicating where to sample to cover the full range of soil variability, thus greatly improving the efficiency of the ground survey. Also, in contrast to a traditional soil map, the EM map shows the pattern of soil variation at a high resolution (at a large scale), which in itself can be very helpful in establishing the vineyard.

In some cases EM data can be correlated with the variation in a single soil property, such as salinity or clay content, and a map of that property produced. Figure 2.6 is an example of a vineyard map in which the variation in the EM signal correlates well with changes in soil clay content. In general, the accuracy of any EM-derived map of a particular soil property depends on how well the EM signal can be calibrated against actual measurements of that property in the field. The next section gives an example of calibration after an EM survey.

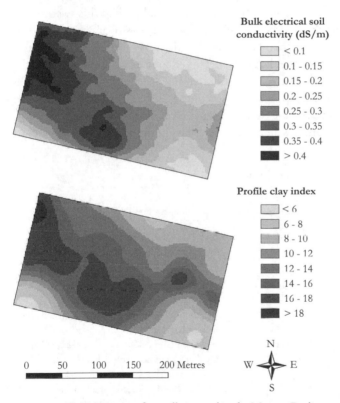

Figure 2.6 (A) EM38 map of a small vineyard in the Murray-Darling region, Victoria, Australia. (B) A map of soil profile clay index derived from 130 sampling points in the same vineyard. The index is the mean of the clay content (expressed as a percentage) at depths of 5 to 15 cm and 45 to 55 cm. (Courtesy of Dr. Robert Bramley, CSIRO Ecosystem Sciences, Adelaide, South Australia.)

Texture Analysis and Calibration for Texture

Texture refers to the "feel" of a soil and depends on the size distribution of the mineral particles. Texture analysis is especially important for obtaining estimates of readily available water, as discussed in chapter 4. In a traditional soil survey carried out by an experienced surveyor, texture is usually assessed by hand-texturing using the procedure outlined in table 2.1. The results can be confirmed by particle-size analysis of selected samples in a laboratory (see the following). However, a laboratory analysis of texture is mandatory for the calibration of an EM survey or any other means of sensing.

Calibration of an EM38 or similar instrument for texture is based on a particle-size analysis of soil samples (the sample set) that represent the full range of texture variation in the vineyard. The samples are collected from specific points so that their EM reading and location are known. Ideally, soil salinity, as measured by the electrical conductivity (*EC*) of a saturated paste, should be less than 0.6 deciSiemens (dS) per meter (dS/m), because salinity higher than this swamps the effect of any other variable on the EM signal. For particle-size analysis, a known weight of dry soil is broken into its constituent particles by mechanical and chemical treatment. Gravel and stones larger than 2 mm diameter are excluded, with the remaining material called "fine earth." Although the distribution of particle sizes in the fine earth is continuous, for practical purposes the distribution is divided into classes. Not all countries use the same size-class divisions, but two of the most widely accepted classification schemes are described in the following.

Table 2.1 A Simplified Guide to Field Texturing

Sample preparation	Texture class	Distinguishing features
Take enough soil to fit into the palm of your hand, slowly moisten it with water and knead it between finger and thumb. Continue kneading and wetting until the soil "bolus" just sticks to your fingers and all aggregates are broken down (may take 1–2 minutes). Assess the texture class by feel, sound, and the length of the soil ribbon that can be squeezed between finger and thumb	Coarse sand	Particles are large enough to be seen and they grate together during kneading; bolus breaks apart[a]
	Fine sand	Particles may be difficult to see and feel but can be heard grating or "squeaking" when the bolus is kneaded close to the ear; bolus lacks cohesion
	Silt	Confers a smooth silkiness to the bolus, which is coherent but not very sticky (particles cannot be felt)
	Clay	The bolus is hard to knead initially but gradually becomes very sticky; forms an intact ribbon up to 50 mm long for a light clay and 75+ mm for a heavy clay[b]

[a] Humified organic matter makes a sandy texture feel smoother and more cohesive.
[b] The type of clay mineral affects stickiness. When thoroughly wet, montmorillonite clay is sticky, but kaolinite clay is not (see "Retention of Nutrients by Clay Minerals and Oxides," chapter 3).

The International Scheme

Clay	<0.002 mm
Silt	0.002–0.02 mm
Fine sand	0.02–0.2 mm
Coarse sand	0.2–2 mm

The United States Department of Agriculture Scheme

Clay	<0.002 mm
Silt	0.002–0.05 mm
Fine sand	0.05–0.1 mm
Medium sand	0.1–0.5 mm
Coarse sand	0.5–1 mm
Very coarse sand	1–2 mm

The amounts of clay, silt, fine sand, and so forth are calculated as a percentage of the fine earth, and their proportions determine the soil's texture. When these proportions are plotted on a triangular diagram, shown in figure 2.7, we can pinpoint the textural class of the soil. For example, a soil with 35% clay, 30% silt, and 35% sand is classed as a clay loam. Experienced surveyors can estimate the approximate proportions of clay, silt, and sand by hand-texturing in the field.

Based on the measured clay contents of the sample set, an EM calibration graph similar to that in figure 2.8 is drawn up. The line of best fit through the points is used to convert EM readings to clay content for the whole survey area and can also be applied to subsequent EM surveys done on this soil type.

Ground-truthing

Calibration of EM readings for clay content is an example of "ground-truthing," which is the general term describing the use of actual soil analyses to check the accuracy of information provided by remote or proximal sensing. The better the calibration, the more accurate the soil map will be in representing the soil property and its variation. As described previously, the samples for analysis can be obtained from soil pits or by augering or coring the soil. When an EM map such as that shown in figure 2.6 is available, the placement of the sampling points can be guided by the variation in EM signal to ensure the maximum range of soil variation is sampled.

After sampling and calibration are complete, the map's precision can be checked by a random sampling of the area to determine whether the location, for

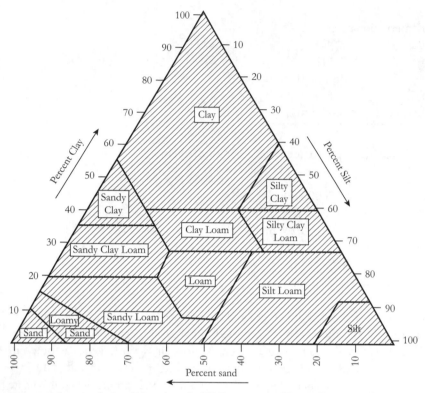

Figure 2.7 A textural triangle based on the US Department of Agriculture particle size classification (similar to that used in Europe). (White 2003; reprinted with permission of the United States Department of Agriculture.)

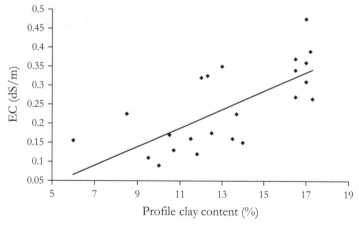

Figure 2.8 Calibration graph for electrical conductivity in terms of profile clay content. Twenty-five locations were selected at random from the EM38 map of figure 2.6A. The line of best fit between electrical conductivity and clay content is shown. (Courtesy of Dr. Robert Bramley, CSIRO Ecosystem Sciences, Adelaide, South Australia.)

example, of changes in clay content or soil depth shown on the map agrees with that found on the ground. Indeed, more intensive soil augering can be used to spot-check the values of other soil properties, such as pH, color, stone content, and profile form, which are based on the relatively few observations made in profile pits.

Describing a Soil Profile

The soil profile should be examined in a backhoe pit, dug to at least 1.5 m or to the parent material if this is shallower. The pit is necessary to reveal properties of the soil in an undisturbed state, especially structure, which cannot be observed by augering. From a deep pit examination, information can be obtained below the sensing depth of ground-penetrating radar, laser-imaging radar, and EM survey. Horizons and layers in a soil profile were introduced in box 1.1, chapter 1, distinguishable by such obvious properties as color, texture, and structure. Descriptions of vineyard soils may refer to horizons or to depth intervals, such as 0–30, 30–60, 60–90, and 90–120 cm. Whichever approach is used, the main focus is on identifying the limitations to successful vine growth, especially with respect to drainage and root penetration, with the latter being more important in dry-grown vineyards.

A survey report is enhanced by the inclusion of color diagrams of the main soil profiles annotated with a description of their important features, as listed in table 2.2, which draws on recommendations made by practicing soil surveyors (e.g., McKenzie, 2000 and Maschmedt et al., 2002). Soil survey field handbooks, such as that of the Soil Survey Division Staff (1993) or the National Committee on Soil and Terrain (2009), give more details of methods.

To make measurements in the field, a surveyor requires the following tool kit:

- A tape measure (preferably metric)
- A strong knife (and/or geological hammer)
- A pH test kit
- A Munsell color chart (or paint chips)
- Distilled water (or rainwater if not available) and plastic cup
- A hand lens (10× magnification)
- A small bottle of 5% hydrochloric acid with dropper

In addition to soil profile descriptions, winegrowers are interested in the fertility of their soils, which involves laboratory testing (see "Soil Testing," chapter 3) and also whether there are potential biological problems (see "Soil Organisms," this chapter).

Table 2.2 Soil Profile Properties Relevant to Site Selection for Vineyards

Key soil property	Relevance of this property	Method of measurement
Soil depth—whole profile or depth to a compacted (impeding) layer	Affects rooting depth and the possibility of waterlogging; may indicate a need for deep ripping	In centimeters; a compacted layer resists insertion of a penetrometer (see "Consistence")
Color	Orange to brick-red colors (resulting from ferric oxides) indicate good drainage; dark colors indicate organic matter; "mottling" indicates periodic waterlogging (see box 1.3, chapter 1)	By eye (important to specify whether soil is wet or dry); describe mottles and background soil color separately
pH	Affects the availability of several nutrients; pH <4 and >9 inhibit root growth	Universal indicator in the field (accurate to ±0.5 pH units) or a pH meter in the laboratory
Texture, stones, and gravel	Influence a soil's water-holding capacity, drainage, structure, and ease of cultivation	Field texturing (table 2.1) or particle-size analysis in the laboratory; proportions of gravel (2–20 mm) and stones (20–60 mm) estimated by eye
Structure, including sodicity and cracking	Influences soil stability when wet, hard-setting when dry; also determines aeration, drainage, and ease of root penetration; montmorillonite (cracking) clays swell when wet and shrink when dry	From a spade sample, assess the degree of aggregation as well as aggregate sizes and shapes; note frequency and width of cracks; check stability of small aggregates placed in distilled water or rainwater (see figure 4.5, chapter 4)
Consistence	Describes the strength and bulk coherence of a soil, which influence drainage, root penetration, and ease of cultivation	Test ease of crushing dry 20–30-mm aggregates between finger and thumb or underfoot—"loose," if very easy, to "rigid" if very hard; measure bulk soil strength with a penetrometer (see "Measuring Soil Strength in the Vineyard," chapter 4)
Presence of limestone or chalk ($CaCO_3$) and its hardness	Indicates neutral to alkaline pH; usually associated with good structure but can restrict rooting if very hard	Drops of dilute hydrochloric acid cause fragments to fizz from CO_2 release; if rock, note whether fractured or not
Rippable rock or impeding subsoil layer	Influences rooting depth and soil drainability; if fractured probably can be ripped	Assess hardness with a geological hammer; estimate the extent of fracturing

Other Factors in Site Selection

Water for Irrigation

An adequate supply of good-quality water is essential for vineyards that need irrigation. In regions of hot, dry summers, such as La Mancha in Spain, the Central Valley of California, or the Murray-Darling and Riverland regions in Australia, irrigation water is essential to produce a satisfactory crop. Indeed, during periods of severe hot, dry weather in Australian inland regions during the "millennium drought" of 1996–2008, the vineyards that survived best were those where the irrigation water was applied at the maximum rate just before and during the stress periods. In cool climate regions, however, irrigation may be only supplemental in that it is used for vine establishment and when unusually dry conditions occur.

Water for irrigation may come from groundwater, surface water, reclaimed water, or a combination of these sources. Where a suitable site exists, a winegrower may construct a farm dam; in other cases, water may be pumped from a river or lake, or supplied by pipe line as in the Barossa Valley region, South Australia. The total water requirement for a vineyard will depend on three things:

1. The climate, especially the balance between rainfall and evaporation and the seasonal distribution of rainfall (see "Evaporation and Transpiration," chapter 4).
2. The likely peak demand as determined by the area to be irrigated, grape variety, planting density, yield objectives, and irrigation method.
3. Additional factors, such as soil storage capacity, whether overhead sprinklers are to be used for frost control, or a winery is to operate onsite.

Given that irrigation in most modern vineyards is applied through surface or subsurface drip lines, the likely peak demand will depend on the number of vines per ha, the number of drippers per vine, and the delivery rate of the drippers. Box 2.3 gives examples of how to calculate the peak delivery rate and total quantity of water required.

Box 2.3 Water Needs for an Irrigated Vineyard

Drippers for surface drip irrigation have delivery rates of 0.6 to 4 L/hour (hr). Delivery rates of 0.8 to 1.2 L/hr are preferred for soils in which water entry is slow. In this case, two or three drippers can be placed in the drip line near each vine for the required amount of water to be applied during an irrigation (figure B2.3.1).

(continued)

Box 2.3 *(continued)*

Figure B2.3.1 Drippers in action in a vine row. Tensiometers are in place to measure soil water suction.

Calculation of peak supply rate is illustrated by an example for midsummer (January) for the Riverland region, South Australia. The vine density is 2020 vines per ha, with two drippers per vine delivering 2 L/hr each. Thus the pump and pipe system needs to deliver

$$2 \times 2 \times 2020 \text{ L/ha/hr} = 8080 \text{ L/ha/hr}$$

Suppose the water required per vine, based on evaporative demand, is 16 L/day. To deliver this quantity of water, the irrigation system should run for 4 hr/day and the total daily water requirement would be

$$8080 \times 4 = 32,320 \text{ L/ha/day} \cong 32,000 \text{ L/ha/day}$$

To keep the vines fully supplied with water in midsummer, the water required per ha for one week would be

$$32,000 \times 7 = 224,000 \text{ L}$$

This figure could be rounded to 250,000 L/ha to allow for leakages and unforeseen circumstances. Irrigation water volumes in Australia are expressed in megaliters (i.e., 1 million liters [ML]) so 250,000 L/ha is equivalent to 0.25 ML/ha. Note also that 1 ML/ha is equivalent to a water depth of 100 mm (compare with rainfall units).

(continued)

Box 2.3 *(continued)*

A comparable figure for water required is obtained for the Riverina region in New South Wales, Australia, for a similar vine density. Note that for higher vine densities, the water requirement increases proportionally. A similar calculation can be done for the Lodi Viticultural Area in the Central Valley region, California (see www. lodiwine.com), except that water volumes are measured in acre-inches. Peak demand occurs in July when a typical vineyard's average daily water need is about 0.33 acre-inches. This requirement amounts to 0.33 inches of water spread over 1 acre of land and is equivalent to 8.4 L/m² (conversion factors from metric to U.S. units and vice versa are given in appendix 1). Thus the total quantity of water required daily per ha is

$$8.4 \times 10,000 \text{ L/ha/day} = 84,000 \text{ L/ha/day}$$

and one week's supply amounts to

$$84,000 \times 7 = 588,000 \text{ L/ha (rounded to 0.6 ML/ha)}$$

However, regulated deficit irrigation (see "Controlling the Soil Water Deficit through Irrigation," chapter 4) is advocated in these summer-dry inland regions to control vine vigor and improve grape quality. This is especially so in the Lodi district where the peak water requirement, especially in the weeks between fruit set and veraison, could be 30% to 65% of the calculated amount, depending on vine density and canopy management. Thus for low-density, vertically trained vineyards under regulated deficit irrigation at Lodi, the amount of water required per ha per week could be as small as

$$0.588 \times 0.3 = 0.176 \text{ ML}$$

or as large as 0.382 ML. Note that if a grass cover crop is growing actively in the mid-rows, the amount of water required could be twice as much as that required by the vines alone.

Water Quality

The most critical factor in water quality is the concentration of dissolved salts—mainly, sodium chloride (NaCl), calcium chloride ($CaCl_2$), and magnesium sulfate ($MgSO_4$). Over time, these salts accumulate in the soil because the water supplied by irrigation is used by plants and evaporated. The salt concentration is most readily measured by the *EC* of the water, usually expressed in units of dS/m. From analyses of many water samples, the following approximate relationship has been developed:

1 (dS/m) = 640 mg/L of total dissolved salts (*TDS*)

Table 2.3 explains the relationships between a number of the expressions used for *EC* and the concentration of salts in water.

Table 2.3 Common Expressions for the *EC* and Salt Concentration of Water

Expression	Units	Numerical equivalence
EC value	deciSiemens (dS)/m	—
EC value	milliSiemens (mS)/cm	Same as dS/m
EC unit	microSiemens (µS)/cm	Multiply dS/m by 1000
Salt concentration	milligram (mg)/L	—
Parts per million	microgram (µg)/mL	Same as mg/L

Note. EC = electrical conductivity.

Because irrigation water is concentrated two- to threefold by evaporation from the soil, water of *EC* more than 0.8 dS/m should be avoided for vines, except where salt-tolerant rootstocks are used (see "Salt, Drought, and pH Tolerance," chapter 5). Less water is lost by evaporation with drip irrigation, so water of higher *EC* can be used than is acceptable for spray or flood/furrow irrigation. Subsurface drip irrigation is even more conservative of water than surface drippers, and therefore water of higher *EC* is potentially suitable. However, salinity and sodicity may build up in the irrigated soil under these conditions, as is discussed in "Leaching, Salinity, and Sodicity Control," chapter 4.

Other analyses recommended for irrigation water are

- pH
- Na, Cl, and boron (B) concentrations, for potential toxicities and effects on wine quality
- Ca, Mg, and carbonate concentrations for assessing water "hardness"
- Iron (Fe) concentration, because if gelatinous Fe precipitates form, they can block drippers and even pipelines

Knowing the concentration of Na, Ca, and Mg ions allows the irrigator to calculate the sodium adsorption ratio (*SAR*) of the water, which is important for the reasons outlined in box 2.4.

Reclaimed Water

Winegrowers may consider using reclaimed water when fresh water is in short supply. Reclaimed water is water from sewage works and processing industries that has been treated to a standard appropriate for its intended use. This type of water is being used successfully in wine regions that are reasonably close to a reliable source, such as in McLaren Vale, South Australia. Because reclaimed water is relatively high in nutrients (the water in McLaren Vale contains approximately 20, 9, and 24 kg/ha of nitrogen (N), phosphorus (P), and potassium (K), respectively, for every ML/ha applied), fertilizer inputs to the vines should be reduced accordingly. However, the

---

Box 2.4 The Sodium Adsorption Ratio of Irrigation Water

Ions are charged forms of the elements that are released when compounds are dissolved in water. The salts of the elements Na, Ca, and Mg in irrigation water are almost completely ionized to produce the cations Na^+, Ca^{2+}, and Mg^{2+} that are electrically balanced by the anions Cl^-, SO_4^{2-}, and HCO_3^-. The *SAR* of the water is defined as

$$SAR = \frac{[Na^+]}{\left[\dfrac{Ca^{2+} + Mg^{2+}}{2}\right]} \tag{B2.4.1}$$

The ion concentrations in this equation, indicated by the square brackets, are measured in millimoles of charge[1] per liter. Through a process of ion exchange, the *SAR* determines the proportion of each cation on clay surfaces, which in turn affects the way that the clay behaves in water (see box 4.3, chapter 4). In soil, the critical factors are the proportion of Na^+ relative to Ca^{2+} plus Mg^{2+} and the total concentration of dissolved salts. The proportion of exchangeable Na^+ relative to Ca^{2+}, Mg^{2+}, and K^+ is usually expressed as the exchangeable sodium percentage. With prolonged use of irrigation water of *SAR* ≥6, the soil exchangeable sodium percentage can rise to a value >6 and the soil becomes "sodic," a condition that predisposes to structural problems. The management of sodic soils is discussed under "How to Improve Soil Drainage" in chapter 4.

[1]Moles of charge are explained in box 3.4, chapter 3.

---

high *TDS* of reclaimed water (1000–1800 mg/L) can create problems after regular use unless the soil is leached by rain or a substantial volume of low-salt irrigation water is applied during winter to remove the excess salts. For example, salts seem to be accumulating in some McLaren Vale vineyards that have been using reclaimed water for a number of years (Arbuckle, 2013). One option to minimize salt build-up is to blend or "shandy" reclaimed water with water of low *TDS*. Given current concern about industrial wastewater polluting supplies of potable water, large wineries have been installing their own treatment plants to treat their wastewater and produce reclaimed water for possible use in the vineyard or for irrigating woodlots and pasture (see "Winery Waste and the Soil," chapter 5).

Soil Organisms

Some pest and disease organisms of grapevines live for part or all of their life cycle in the soil, for example, the root aphid phylloxera and parasitic nematodes. The history of a site is especially relevant in indicating what tests should be carried out.

In Australia, a vineyard area can be classed as a phylloxera infested zone, a phylloxera risk zone (where phylloxera may exist but the area has not been surveyed), or a phylloxera exclusion zone (where no phylloxera has been found in surveys). Many vegetable crops such as tomatoes, potatoes, and carrots, as well as tree crops such as almonds, are hosts for nematodes that can attack and damage vine roots. Hence the soil should be tested for the density and types of nematode present. Chapter 5 describes these organisms in more detail, as well as some of the soil-borne pathogenic fungi and bacteria that attack grapevines (see "Soil Interaction with Fungal and Bacterial Diseases" in chapter 5). Certain soil and vegetation factors predispose to biological problems. Phylloxera damage is generally more severe in clay and heavy clay soils than sandy soils, whereas nematode damage is more severe in sandy soils.

In testing for soil pests and disease, samples of moist soil should be collected by spade or auger from several locations, bulked, and sealed in polythene bags to be sent to a qualified laboratory for analysis. A bulk sample should represent no more than 1 to 2 ha. Testing for nematode types and numbers requires moist soil that is kept as cool as possible until analyzed. If unhealthy vines or other plants are seen onsite, separate samples should be taken where these occur and also in apparently healthy areas.

The presence or absence of phylloxera and/or nematodes will determine whether *Vitis vinifera* cultivars on "own roots" or on rootstocks should be planted. All *V. vinifera* cultivars are susceptible to phylloxera, as are any rootstocks that have *V. vinifera* stock in one parental line. Thus although own-rooted cuttings are cheaper to buy, resistant rootstocks are the only sure way to avoid serious damage from phylloxera. Chapter 5 discusses the advantages and disadvantages of rootstocks. Regardless of whether rootstocks are chosen, the cuttings or "rootlings" (rooted cuttings) should be free from phylloxera, nematodes, and any potential fungal, bacterial, or viral pathogens (see box 2.5).

Box 2.5 Preplanting Treatment of Rootlings and Cuttings

Hot-water treatment is essential to eliminate nematodes and phylloxera on the roots of dormant rootlings and to control viruses, bacteria, and fungi that are potential pathogens in dormant cuttings. The cuttings must be fully dormant before treatment.

The treatment is as follows:

Rootstocks and *V. vinifera* cuttings   Hot water at 50°C for 30 minutes
Rootlings (bare roots)   Hot water at 50°C for 5 minutes

Bundles of cuttings and rootlings should be hydrated and brought to ambient temperature before the hot-water treatment. Immediately after treatment, the bundles are plunged into clean water at ambient temperature for 20 to 30 minutes. Those that are to be kept in cold storage before being used may be dipped in a fungicide such as Chinosol (8-hydroxyquinoline sulfate) to control pathogens.

Slope, Aspect, Nearness to Water, and Heat Input

Slope combined with aspect is important for maximizing the amount of solar radiation received by vineyards at higher latitudes (between 40° and 50°N or S). As shown in figure 4.16, chapter 4, part of the radiation absorbed at the land surface warms the soil. The sun's rays pass through a greater thickness of the earth's atmosphere at high latitudes than at the equator, which means the intensity of sunlight received by soil and vegetation is less at these latitudes. The decrease in sunlight intensity is greater during autumn through winter into spring. However, because steep slopes that face toward the equator receive sunlight at a larger angle of incidence than flat land, their potential for absorbing radiant energy and soil warming is greater. Figure 2.9 illustrates these effects.

For these reasons, the most favored sites in cool-climate regions are predominantly south-facing slopes in the northern hemisphere and north-facing slopes in the southern hemisphere. However, this generalization needs to be modified for particular regions, such as in the Côte d'Or in Burgundy, where, because of the local topography, many of the slopes are east-facing. Unusually, in the cool Rhinegau region of Germany, west-facing slopes are preferred because during the critical autumn ripening period, morning fogs are common, which usually clear by afternoon to the benefit of vines on west and southwestern slopes.

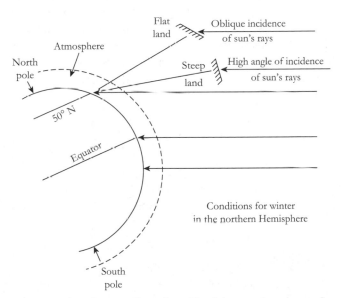

Figure 2.9 The influence of latitude and land slope on the amount of solar radiation received and absorbed at earth's surface (White 2003).

Although the duration and intensity of sunlight are not limiting in hotter regions, such as in La Mancha, Spain, the Central Valley of California, and the Murray-Darling region of Australia, excessive exposure of fruit to direct radiation can cause sunburn and too rapid a rate of ripening. This problem can be managed through row orientation and row spacing and by allowing more canopy shading. Row orientation and tree belts are important for protection from wind damage in some maritime areas (e.g., in many vineyard areas of New Zealand's North Island and along the central and northern coast of California).

Slopes also induce cold air drainage, which helps to avoid damage to vines on the slopes from low temperatures in winter and frost in spring. Most grape cultivars require at least 180 frost-free days, and frosts after bud burst can cause serious crop losses. In undulating topography, cold air collects in valley bottoms, whereas warm air rises. In mountain regions, such as central Virginia, the mid-slopes are best for vines because cold air drainage makes the lower slopes and valley bottoms frost-susceptible in spring. However, vines on the highest slopes are more likely to suffer winter freeze damage (at temperatures of 2°–15°C below zero), depending on how well the vine tissue is acclimatized.

The movement of air of different temperatures (and hence density) can make up a convection cell (figure 2.10). These cells also develop near large bodies of water, such as lakes, large rivers, and estuaries. Because the specific heat capacity of water is much greater than soil, a large water body is a greater heat sink than nearby land, so that at night warmer air over the water rises, to be replaced by cooler air draining from the land. The moderating effect of these convection cells means that vineyards near large water bodies do not suffer the extremes of temperature, either diurnal or seasonal, which occur in inland (continental) climates. Classic examples of this effect occur in the Médoc region, France, which is bordered by the Gironde estuary to the east and the Bay of Biscay to the west, the Margaret River region of Western Australia, and the Mornington Peninsula region, Victoria, Australia.

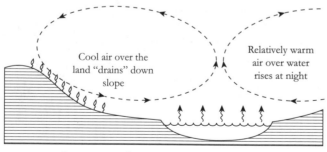

Figure 2.10 Convection cells on a clear night for vineyards near a large body of water
(White 2003).

Modifying the Soil–Vine Environment

Site Potential

Each site has an inherent potential that reflects its suitability for particular varieties and the likely vigor (relating to vegetative growth) of the vines to be grown. Given that site potential is determined by geology, soil, topography, climate, and human influence (if the land form is reconstructed by earth-moving), we can summarize the broad differences between low- and high-potential sites using a combination of the soil properties given in table 2.2 and additional geological factors, as shown in table 2.4.

Additionally, topography influences site potential through soil depth, drainage, temperature, and aspect. Soil on steep slopes is prone to erosion and therefore tends to be shallower than soil on gentle slopes and in valley bottoms. Soil at the top of slopes is usually better drained than at the bottom. Temperature and aspect effects were discussed earlier, but for a given suite of cultivars and scion–rootstock combinations, the interaction between rainfall and soil has a powerful influence on site potential, as expressed through vine vigor. Deep, well-drained soils that are naturally fertile and well watered will predispose to "excess vigor."

Table 2.4 **Soil and Geological Features of Low- and High-Potential Site[a]**

Low-potential sites	High-potential sites
Often on conglomerates, sandstones, or shales and their metamorphic products, in which little weatherable mineral remains	Often on igneous rocks, especially basalt, diabase, and dolerite, relatively unweathered shales, their metamorphic products, and alluvium derived from these rocks
Underlying rock is hard and unrippable	Underlying rock is soft and rippable; no shallow barriers to root growth
Shallow soil (<0.5 m)	Deep soil (>1 m)
Sand to sandy loam, or a high proportion of gravel and stones; small water-holding capacity	Clay loam to light clay and <5% stones; large water-holding capacity
Low pH and exchangeable Al, or high pH and salinity	pH 5.5–7.5 and no exchangeable Al or salinity
Little organic matter, usually a shallow topsoil and low mineral nitrogen supply	Plenty of organic matter, dark A1 horizon that can be more than 20 cm thick; potentially high rate of mineralization of organic nitrogen
Weak soil structure, especially in the subsoil; aggregates unstable in water	Well-aggregated soil; firm aggregates stable in water
Poorly drained subsoil (mottled or blue-gray colors throughout)	Well-drained soil profile (uniform orange to brick-red colors in subsoil and no mottling)

[a] Varieties appropriate for high- and low-potential sites are discussed in chapter 5.

Excess vigor reflects a lack of balance between the vegetative and reproductive stages of a vine's development. When too much vegetative growth occurs, especially between flowering and veraison, fruit set can be poor and shading of the fruit can lead to slow and uneven ripening, which affects the quality of the harvested fruit. Even in dry regions, irrigated vines on deep, fertile soils can suffer from excess vigor if too much water is supplied. Examples occur in the Murray-Darling and Riverina regions in Australia and on alluvial soils of the Napa and Central Valley regions of California. Chapters 3 and 4 discuss further the problem of excess vigor induced by too much N and water, respectively.

Preparing the Site by Cultivation and Ripping

Soil treatment starts in summer. If the land has been cleared recently of trees or old vines, all residues must be removed or burned and the soil cultivated or ripped to bring up old roots that may harbor pest and disease organisms. Deep cultivation to about 0.5 m in depth can be done with a moldboard plow, which turns a furrow slice. Although effective in bringing up old roots, moldboarding has the disadvantage of causing deep soil compaction, especially when the soil is wet, through pressure from the plowshare and tractor wheels in the furrow. In calcareous silty-sandy soils in the Hérault region, southern France, compaction of this kind has caused localized stunting and withering of vines and yield decline (figure 2.11). Consequently, winegrowers

Figure 2.11 Stunted unhealthy vines as a result of compaction in silty, sandy soils in the Hérault region, Languedoc-Rousillon, France.

have turned to deep ripping (sometimes called subsoiling), which involves no soil inversion, and compaction under the tractor wheels is broken up by the ripper shafts that follow. However, old roots are not brought to the surface, so the soil may need to be fumigated to destroy nematodes (see "Preplanting" in this chapter).

Deep ripping is often used to break up a compacted layer, either induced or formed naturally, which restricts root development and seriously impedes drainage. For example, large rippers drawn by crawler tractors are used to rip shallow soils in the Napa region of California. In the Western Cape region of South Africa, a "finger delve" plow is used to break up compacted B horizons in duplex soils, without inverting the soil (figure 2.12). With this plow, as with rippers, lime can be incorporated at depth to raise the pH and improve soil structure; or in soils where acidity is not a problem, gypsum can be added to improve structure.

In established vineyards, compaction from traffic in the mid-rows can be serious, more so in sandy soils; but this, too, can be alleviated by ripping. South African scientists have developed a novel cultivator for deep disturbance of the mid-row soil to alleviate compaction (figure 2.13). This implement also prunes the roots, which stimulates root growth and rejuvenates old vines of declining

Figure 2.12 A "finger delve" plow used for soil preparation for vineyards in the Western Cape region, South Africa. The maximum depth of ripping is approximately 1.5 m. (Courtesy of Dr. Eduard Hoffman, Stellenbosch University, South Africa.)

Figure 2.13 A cultivator used for alleviating compaction in the mid-rows of vineyards in the Western Cape region, South Africa. (Courtesy of Dr. Eduard Hoffman, Stellenbosch University, South Africa.)

vigor. Many duplex soils in Australia and South Africa have compacted A2 horizons, which can be improved by mixing the A2 with the organic-rich A1 horizon.

Box 2.6 outlines the methods and soil conditions for successful ripping to alleviate restrictions on rooting and allow modification of subsoil drainage and chemical state. The conditions to be alleviated must impose a serious constraint on vine growth; otherwise deep ripping may simply encourage vines to grow too vigorously, which is usually not conducive to producing high-quality fruit.

Erosion Control

After deep cultivation or ripping, the soil surface is leveled to prepare for planting. Bare soil on slopes is vulnerable to erosion, especially when the vines are dormant in winter, so some form of erosion control should be used. On gentle to moderate slopes (0%–5%), a cover crop can be used to protect the soil from erosion and cultivated into the soil before the vines are planted. Suitable cover crops are discussed in "Cover Crops and Mulches," chapter 5. On steeper slopes (6%–15%), earthworks

Box 2.6 Methods of Deep Ripping

To achieve shattering of a restricting layer or a dense subsoil, deep ripping is best done in late summer to autumn. Ideally the soil water deficit (see "How a Soil Water Deficit Develops," chapter 4) should be about 50 mm in a sand or 100 mm in a clay—this corresponds to a friable soil consistence when the soil aggregates will crumble under light to moderate pressure (see "Soil Strength," chapter 4). The rationale for this condition is that the soil should not smear but shatter as the ripper blade passes through. In some cases it may be necessary to sow a winter cereal, to be harvested in the summer, to dry the soil profile before deep ripping. Rip to a 1-m depth if possible. Achieving this depth requires a 1.5 m-shank and a D8 or similar tractor. The subsoiler blade and its angle of strike are chosen according to the problem to be alleviated, as shown in table B2.6.1. If the soil is ripped at the correct water content (a friable consistence) at a 2-m spacing (with wings attached), the fissures created by the blade should intersect at the soil surface. This is important for good drainage.

Table B2.6.1 Recommendations for Ripping Subsoils

Soil problem	Blade type	Rake angle[a]
Hard rock	Do not rip	
Weathered rock	Rip with a wingless blade	
Cemented layer in profile	Rip with a wingless blade	
Clay subsoil	Rip with a winged blade	20 degrees from horizontal
Compacted sands	Rip with a 90-degree point	

[a] The angle of the blade is critical for proper lift and shattering of the soil, with minimal lateral compaction.

Adapted from Cass et al. (1998).

such as contour banks or terraces are required to break up long slopes and divert runoff water laterally, preferably to a grassed waterway at the edge of the block.

Old, established vineyards in Europe, such as the Hill of Hermitage in the Rhone Valley, France, have broad terraces on which vines are planted in rows up and down the slope (figure 2.14). In newer vineyards, the terraces are much narrower so that vine rows must follow the line of the terrace across the slope, as seen in the Friuli region of northern Italy (figure 2.15). Usually two or three rows of vines are planted on each terrace, to make mechanical operations in the vineyard easier. Although many older vineyards in the Mosel and Rhine Valleys of Germany have been planted on very steep slopes, planting on slopes greater than 15% is not generally recommended, except in the Vaud and Valais regions

Figure 2.14 Old vines on broad terraces on the granitic Hill of Hermitage, Rhone Valley region, France.

Figure 2.15 Terraces in a vineyard in the Friuli region, northern Italy.

of Switzerland, for example, where there is limited land for vineyards that is not very steep (see figure 1.5, chapter 1).

Preplanting

In land previously under vines, there are likely to be pests and disease that should be controlled before new vines are planted, as discussed earlier in "Soil Organisms." For example, in parts of California, nematodes and oak root fungus are problems where old roots remain buried. The pathogenic fungus *Phytophthora* can also be present. Since the phasing out of the ozone-depleting fumigant methyl bromide, products such as Vapam HL (sodium N-methyldithiocarbamate) and Telone (1,3 dichloropropene) are being used to suppress or control nematodes and soil-borne diseases. The websites www.amvac-chemical.com and www.dowagro.com/soil/ give further details for the use of these fumigants. Fumigation is most effective when the soil surface is covered or packed down by rolling after the fumigant has been injected. Because fumigants kill beneficial soil organisms as well as pests, a preferred option for nematode control is to use nematode-resistant rootstocks (see "Biological Control in Vineyards," chapter 5), but this will not necessarily control soil-borne pathogens.

Legume-based cover crops plowed-in before planting stimulate beneficial soil organisms, helping young vines resist pest and disease attack until they are well established. Fertilizers can be applied when the cover crop is sown, but rock phosphate, lime, or gypsum should be incorporated into the soil when it is ripped or plowed, because their beneficial effects depend on coming into contact with as much soil as possible. If the topsoil is shallow, soil may be mounded or "hilled" along the rows to increase its depth and improve drainage, taking care that when rootstocks are used the graft union is not covered. Mounding soil around the vine trunks in winter helps to protect the vines in regions where severe freezing can occur.

Vineyard Layout and Vine Spacing

Vine spacing should be chosen according to site potential and grape variety to satisfy the dual objective of high-quality fruit from each vine and optimum yield per ha. Historically, in European vineyards, vines have been planted on relatively poor soils on slopes, leaving the more fertile soils of river valleys and flood plains for the cultivation of essential food crops. In maritime, and higher-rainfall continental regions (e.g., Bordeaux and Burgundy, respectively), vine rows could be closely spaced (1 m) with close in-row spacings (1 m) so that fruit yields per ha are optimized. Such close spacings accommodated the introduction of vertically shoot positioned trellising, which improved leaf exposure to sunlight, provided that the vine rows were hedged to prevent too much shading. Figure 2.16 shows an example from the Médoc region of France where fit-for-purpose machinery is employed for viticultural operations. In more arid regions, as in parts of central

Figure 2.16 Vines at 1-m spacing on sandy, gravelly soil at Chateau Giscours, Margaux Appellation, Médoc region, France.

and southern Spain, vines are planted at wide spacings and pruned in a way that trellis support is not needed.

In addition to the natural factors determining site potential (summarized in table 2.4), vine vigor is influenced by whether irrigation water is used and how much. Thus, on high-potential sites in warm regions under irrigation, wide between-row (up to 3.4 m) and in-row spacings (up to 2.4 m) were employed to accommodate the high vigor of the vines grown. Additionally, given that labor was scarce and costly in many New World vineyards, such wide spacings made a vineyard easier to manage with machinery. In the Central Valley region of California, the standard trellis used to support vigorous vine growth was a two-wire, vertical, non-shoot positioned system called the Californian Sprawl (Dokoozlian, 2009). Since the early 1990s, however, even with the Californian Sprawl, row spacings have decreased to 2.6 m and in-row spacing to 1.6 to 1.8 m, giving plant densities of 2200 to 2400 per ha. In coastal regions of California, the Napa and Sonoma Valleys, and most of the cool climate vineyards in Australia and New Zealand, closer vine spacings with vertically shoot positioned trellising are preferred. However, the need to mechanize operations as much as possible to reduce costs generally imposes a limit on the closeness of rows, and achieving an optimum leaf area per meter of row determines the in-row spacing. Nevertheless, some high-end producers in these countries have planted at very high densities, as discussed in box 2.7.

Box 2.7 Examples of High-Density Planting in Some Australian Vineyards

Influenced perhaps by the high planting densities (10,000 vines per ha) in Grand Cru and Premier Cru vineyards of the Côte d'Or region, France, some Australian winegrowers have established vineyards at similar densities, even on high-potential sites. Often these producers will also use low-input organic and/or biodynamic methods. An example in figure B2.7.1 shows own-rooted Pinot Noir vines growing on a black earth derived from basalt in the Geelong region, Victoria, Australia. Low nutrient inputs, minimal irrigation, and a complete grass-clover crop subdue the vigor of these vines. Young vineyards of similar style are also found in the Yarra Valley region, Victoria.

Figure B2.7.1 Pinot Noir vines in close spacing (9000 per ha) in the Geelong region, Victoria.

Another example is found in a vineyard in the Gippsland region, Victoria (receiving approximately 1000 mm rainfall annually), on a fertile basaltic soil, where the normal planting density is 9000 vines per ha, but there is also an experimental block planted at 17,000 per ha! Organic/biodynamic methods are used, without irrigation, but to maintain yields per vine as low as a few hundred grams the vines must be hedged about 10 times per season. Obviously, yields are low (2.5–3.3 t/ha) and the cost of production is high, so it is economically viable only if the wine produced commands a very high price.

In summary, Carbonneau (2009) maintains that tradition has tended to determine vine spacing, with close planting the norm in Old World vineyards, except in very dry areas, and wider spacings used in the larger-scale, more vigorous vineyards of the New World. French Appellation d'Origine Contrôlée regulations, and those of comparable organizations in other Old World countries, limit the quantity of wine produced per ha. Whereas such restrictions on fruit or wine yields were not previously imposed in New World countries (e.g., Australia), in recent years many wineries have imposed limits on the tonnes of fruit per ha. At the same time, planting densities have increased, which has generally resulted in more efficient exploitation of the soil by the vines. According to a winegrower's objectives, vine vigor and yield are now more effectively controlled by choice of rootstock, low input of nutrients and water, use of cover crops, and canopy management (particularly pruning), the last of which is the subject of the book *Sunlight into Wine* by Smart and Robinson, first published in 1991, with the last reprinting in 2013.

Irrigation and Drainage Design

In France, irrigation cannot normally be used if the wine produced is to be marketed under an Appellation d'Origine Contrôlée label. In other countries, winegrowers who want to produce *terroir*-type wines at the premium end of the market will tend to use irrigation sparingly or not at all. However, in many regions of Spain, Australia, South Africa, Chile, and California, where the summers are hot and dry, irrigation has become an essential part of vineyard management. Surface or subsurface drip irrigation is preferred to overhead spray or furrow irrigation because it permits greater control over the water applied and minimizes water loss through evaporation (see "Irrigation Methods," chapter 4).

Vine water use can be made more efficient if the application of water takes account of soil properties such as depth, drainability, and water-holding capacity (see table 2.2). Although organic matter can increase a soil's water-holding capacity, its effect is confined to the topsoil and is more significant in sandy soils than clay soils. For example, Kay et al. (1997) used pedotransfer functions to show that an increase in soil organic carbon (say, from 0.7% to 1.7%) would increase plant-available water only by 2 to 4 mm per 10 cm depth of soil. The influence of various factors on plant-available water is discussed more fully in chapter 4.

If a soil survey reveals zones where one or more of the critical soil properties are substantially different, the irrigation design should be adjusted through the spacing of drippers and their delivery rate (see box 2.3). Pressure self-compensating drippers allow water output per dripper to be as uniform as possible along a pipeline and also over undulating ground. Keeping the pipe system as full as possible between irrigation sessions is important to reduce filling time, as is regular flushing and checks on operating pressure.

One of the functions of deep ripping is to improve soil drainage. However, if poor drainage is due to an impermeable subsoil or a groundwater table that rises within 2 m of the soil surface, an artificial drainage system may need to be installed. Soil conditions predisposing to poor drainage and the means of alleviating this problem are discussed in "What Causes Poor Drainage?" in chapter 4.

Summary Points

1. Site potential expresses the suitability of a site for particular grape varieties and the expectation of vine vigor expressed through vegetative growth. Climate, topography, geology, and soil type all influence site potential and hence site selection for a vineyard.

2. Traditionally, most attention has focused on the regional climate in site selection, and a number of climatic indices have been developed. Locally, climate interacts with topography through land height, slope, and aspect (the mesoclimate), while the nearness to large bodies of water moderates daily and seasonal temperature variations.

3. A more complete approach to site selection involves taking account of other determining factors, notably soil, to develop a local site index (LSI).

4. Because soil is important in determining site potential, detailed knowledge of soil properties, spatially mapped through a soil survey, is an essential prerequisite for the establishment and subsequent management of a vineyard.

5. Soil survey is greatly aided by remote and proximal sensing techniques, such as the EM38, which responds to changes in electrical conductivity (*EC*) down to a depth of at least 50 cm. When the sensed positions are located with a global positioning system (GPS) and the output is streamed to a geographic information system, an accurate map of soil variation can be produced. With such a map, points covering the full range of soil variation can be efficiently located on the ground and examined in soil pits. Samples from the pits can be used to calibrate maps for particular soil properties such as depth, clay content, and salinity.

6. A range of tests is recommended for soil chemical fertility (see chapter 3), physical properties (see chapter 4), and biological condition, including the presence of pest and disease organisms (see chapter 5). From these tests, a winegrower may decide on preplanting treatments such as deep ripping (to break up compacted layers), fumigation (for disease control), terracing (for erosion control), or sowing a cover crop. Soil amendments,

such as lime and gypsum, and rock phosphate fertilizer can be incorporated at this stage and a decision made on the use of rootstocks.

7. Climate and soil factors may determine that full-scale irrigation, or only supplemental watering, is required. If irrigation is to be used, the design should be based on the soil types present and their distribution. The quantity of water required at peak demand can be estimated and water quality assessed, with particular emphasis on total dissolved salts (*TDS*) and dissolved elements such as Na, Cl, Fe, and B. If reclaimed water of high *TDS* is to be used, provision for salt leaching and offsite drainage must be considered.

8. Site potential and the use (or not) of irrigation have influenced vine spacing. In Old World vineyards on poorer soils in higher rainfall regions, planting densities can be as high as 10,000 vines per ha, whereas in New World regions on more fertile soils with full irrigation, lower densities of 1800 to 2200 vines per ha are common. However, higher plant densities are now being adopted, especially in cool climate vineyards of the New World, when the emphasis is more on wine quality than yield.

3

The Nutrition of Grapevines

The Essential Elements

Grapevines must have 16 of the 118 known elements to grow normally, flower, and produce fruit. These essential elements, listed in table 3.1, are also called nutrients and as such are divided into

- Macronutrients, which are required in relatively large concentrations
- Micronutrients, which are required in smaller concentrations

Box 3.1 discusses the different ways of calculating nutrient concentrations in soil, plants, and liquid. Vines draw most of their nutrients from the soil, and so table 3.1 also shows the common ionic form of each element in soil.

Ions, the charged forms of elements, are introduced in box 2.4, chapter 2. For example, carbonic acid (H_2CO_3), which is a compound of carbon (C), hydrogen (H), and oxygen (O), dissociates in water into the ions H^+ and HCO_3^-. This is a chemical reaction that can be written in shorthand form as

$$H_2CO_3 \leftrightarrow H^+ + HCO_3^-$$

The double arrow shows that the reaction can go either forward (to the right) or backward (to the left), depending on the concentrations[1] of H^+ and HCO_3^- relative

---

[1] The concentration of H^+ ions in a solution at pH 5 is 10^{-5} molar (M), equivalent to one H^+ ion per 100,000 molecules of water, which is 10 times greater than the concentration of H^+ ions at pH 6.

Table 3.1 **Macro- and Micronutrients, their Chemical Symbols, and Common Ionic Forms in Soil**

Macronutrient (>1000 mg/kg)[a]	Common ionic forms in soil	Micronutrient (<1000 mg/kg)[a]	Common ionic forms in soil
Carbon (C)	HCO_3^-, CO_3^{2-}	Iron (Fe)	Fe^{3+} (sometimes Fe^{2+})
Hydrogen (H)	H^+	Manganese (Mn)	Mn^{4+} (sometimes Mn^{2+})
Oxygen (O)	H_2O and many ions, (e.g., OH^-, NO_3^-, SO_4^{2-})	Zinc (Zn)	Zn^{2+}
Nitrogen (N)	NH_4^+, NO_3^-	Copper (Cu)	Cu^{2+}
Phosphorus (P)	$H_2PO_4^-$, HPO_4^{2-}	Boron (B)	H_3BO_3, $B(OH)_4^-$
Sulfur (S)	SO_4^{2-}	Molybdenum (Mo)	MoO_4^{2-}
Calcium (Ca)	Ca^{2+}		
Magnesium (Mg)	Mg^{2+}		
Potassium (K)	K^+		
Chlorine (Cl)	Cl^-		

[a] See box 3.1 for units of measurement.

Box 3.1 Units for Nutrient Concentrations and Amounts

Concentration (symbol C)[a] is the amount of a substance per unit volume or unit weight of soil, plant material, or liquid. For example, the concentration C of the element nitrogen (N) can be expressed as micrograms (µg) of N per gram of soil[b], noting that

$$1 \ \mu g \ N/g = 1 \ mg \ N/kg = 1 \ part \ per \ million \ (ppm \ N) \qquad (B3.1.1)$$

An amount is the product of concentration and weight. For example, the total amount of N of concentration C (measured in µg/g) in a soil sample of 100 g is

$$100C \ \mu g \ or \ 0.1C \ mg \qquad (B3.1.2)$$

Because all soil and plant materials contain some water, analyses are best expressed in terms of oven-dry (o.d.) weights. The o.d. weight of a soil sample is obtained by drying it to a constant weight at 105°C; for plant material the drying temperature is 70°C.

The amount of a nutrient is often expressed per hectare (ha) of vineyard. In this case, we need to calculate the weight of soil in 1 ha to a chosen depth, usually 0.15 m (15 cm). Unless it is measured directly, assume the bulk density for the top 0.15 m soil to be 1.33 megagram (Mg) per m^3, so that the weight of dry soil per ha is 1995 Mg or approximately 2000 Mg (2,000,000 kg). Thus the conversion

(continued)

Box 3.1 *(continued)*

of a laboratory result for "available" P in the top 0.15 m of soil, C_p (milligrams per kilogram) to an amount per ha per 0.15 m depth becomes

$$\text{Amount of P per ha} \cong 2C_p \text{ kg P/ha.0.15 m} \qquad \text{(B3.1.3)}$$

If the soil depth of interest is more than 0.15 m, as it would be for mineral N, the conversion factor is correspondingly greater.

Element concentrations are sometimes expressed as a percentage of the dry weight (g/100 g). For example, a soil N concentration of C percent converts to an amount per hectare as

$$C\% = \frac{C\text{g}}{100 \text{ g soil}} = \frac{0.01C \text{ kg}}{\text{kg soil}} \qquad \text{(B3.1.4)}$$

$$\text{Amount of N per ha} = \frac{0.01C \text{ kg}}{\text{kg soil}} \times 2,000,000 \qquad \text{(B3.1.5)}$$
$$= 20,000C \text{ kg N/ha.0.15 m}$$

Analyses of manures and biosolids are often quoted as a percentage of the element in the fresh or "wet" weight. To convert to a dry matter (DM) basis, divide the element concentration in the fresh weight (C_f percent) by the percentage of DM in the fresh weight as follows:

$$C\% \text{ (DM)} = \frac{C_f}{100 \text{ g fresh weight}} \times \frac{100 \text{ g fresh weight}}{\text{g DM}} \times 100 \qquad \text{(B3.1.6)}$$

[a] The concentration C of an element is an example of a variable. Symbols for variables are written in italics.

[b] Units of measurement and their abbreviations are given in appendix 1.

to H_2CO_3 and the strength of this acid. Carbonic acid is a weak acid because it does not completely dissociate into H^+ and HCO_3^- in solution. Ions that carry one or more positive charges (+) are called cations, and those carrying one or more negative charges (−) are called anions.

Sources of the Elements

In sunlight, vine leaves and green stems absorb C and O_2 as carbon dioxide (CO_2) from the atmosphere for the synthesis of sugars. Oxygen and H are also supplied in water (H_2O) derived from the atmosphere and soil. Chlorine (Cl) is abundant as the Cl⁻ ion in the air and oceans. Sea spray containing Cl, sodium (Na), magnesium (Mg), calcium (Ca), and sulfur (S) forms an aerosol that is carried inland, where these elements are deposited in rain or as dust particles (called "dry deposition"). Nitrogen (N) as dinitrogen (N_2) gas can enter soil–plant systems through biological fixation, as described in box 5.2, chapter 5. Small amounts of mineral N are deposited in rain and as dry deposition.

However, in the absence of added fertilizers, the major source of the essential elements is a variety of minerals in the soil and parent rocks that decompose by weathering to release the macronutrients—phosphorus (P), S, Ca, Mg, potassium (K), and Cl, as well as the micronutrients—iron (Fe), manganese (Mn), zinc (Zn), copper (Cu), boron (B), and molybdenum (Mo). Additionally, rock weathering supplies aluminum (Al), silicon (Si), and other elements essential to animals, such as selenium (Se), iodine (I), and cobalt (Co). Yet others that are toxic to humans and animals if they occur in too high a concentration are also released, such as arsenic (As), mercury (Hg), cadmium (Cd), lead (Pb), chromium (Cr), and nickel (Ni).

Excluding Al and Si, the elements from Mn through to Ni normally occur at concentrations of less than 1000 mg/kg in soil and are called trace elements (Fe is the one micronutrient that is not a trace element). However, all these trace elements are potentially detrimental to vines if their concentrations exceed a threshold value, at which point they may be considered contaminants. For example, the high Ni content of some soils formed on serpentine rock at the northern end of the Napa region, California, can cause problems for grapevines, and high Cu concentrations in vineyard soils treated over a long period with Bordeaux mixture can be detrimental to soil organisms. In Australia, the responsible authority for setting out contaminant guidelines is the Standing Council on Environment and Water (www.scew.gov.au/home), which incorporates the National Environment Protection Council.

Supplying Nutrient Ions to Vine Roots

Because most of the essential nutrients for a grapevine are obtained from the soil through the root system, it is important to review factors influencing vine rooting and the soil processes that determine the availability of nutrients to a vine.

The Vine Root System

The grapevine is a perennial plant that has evolved to explore a large soil volume at a low root density (expressed as the length of root per volume of soil). When soil conditions are favorable, the grapevine develops an extensive root system laterally and vertically. For example, based on observations of more than 200 vine root profiles, Smart et al. (2006) in California found that roots normally extend up to 1.5 m laterally from the trunk and that about one-fifth of the total root length is found below 1 m depth. The thicker lateral roots in the top 30 cm of soil do not increase much in number after the third year from planting, but smaller "feeder" roots continue to grow horizontally and vertically from the main framework. Figure 3.1 shows healthy vine roots exploring a well-drained soil.

Soil organic matter is mainly confined to the top 10 to 20 cm of a soil profile and little of any N, P, and S mineralized from organic matter is taken up from depths below 50 cm. However, nutrients such as Ca, Mg, Fe, and Mn can be acquired from rock weathering much deeper in the profile. Some geologists, such as Swinchatt and Howell (2004) for example, believe that the differential uptake of mineral nutrients from depth explains the distinctive taste of wines produced from old vines on different geological formations. Entrepreneurial producers, encouraged by some wine writers, often invoke this kind of explanation for the distinctive *terroir* underpinning their wines, but this explanation has not been confirmed scientifically.

Although repeated cultivation of the mid-rows inhibits root growth in the top 10 cm of soil, occasional cultivation will prune the roots and stimulate root growth. Where there is a subsoil obstruction, such as the dense B horizon of many duplex soils in southeastern Australia, less than 5% of vine roots may

Figure 3.1 Healthy lateral vine roots down to 50 cm in a well-drained red loam soil in the Barossa Valley region, South Australia.

penetrate below 50 cm. Similarly, figure 3.2 shows that most of the vine roots in a soil formed on chalky limestone are confined to a shallow depth above the limestone (unless the limestone is fractured). Deep rooting is also discouraged in irrigated vineyards if the irrigation strategy is to control water availability in the top 40 to 60 cm of soil only (see "Controlling the Soil Water Deficit Through Irrigation," chapter 4).

The Absorbing Root

Vines absorb nutrients dissolved in water, called the soil solution, which is in contact with the roots. Thus a nutrient such as N, which exists mainly in organic combination, must be released into the soil solution as an inorganic ion—ammonium (NH_4^+) or nitrate (NO_3^-)—to be absorbed by a root. From extensive studies of higher plants, it is known that the zone of active absorption is about 20 mm long, located a few millimeters back from the root tip, where there are root hairs (fine outgrowths from the epidermal cells), as shown in figure 3.3. Root hairs greatly increase the absorbing surface so that uptake of a nutrient may exceed its rate of replenishment at the root surface, in which case the concentration around the root decreases and a depletion zone develops.

Figure 3.2 Darkly stained fine roots confined to a shallow Terra Rossa over chalky limestone in the Barossa Valley region, South Australia.

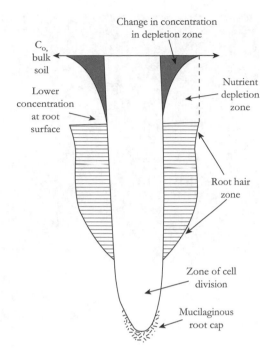

Figure 3.3 Diagram of a grapevine root showing root hairs and the nutrient depletion zone (White, 2003).

Depletion zones most commonly occur with P, K, Fe, Mn, Zn, and Cu because the solution concentrations of these elements are usually low. The ions cross the depletion zone mainly by diffusion, which is a slow process but nonetheless important in maintaining a supply to the root. On the other hand, the solution concentrations of Ca^{2+}, Mg^{2+}, NO_3^-, Cl^-, and sulfate (SO_4^{2-}) are normally large enough for root demand to be satisfied by the amounts swept along in water flowing to the root (the "transpiration stream"), so that no depletion zone develops. Overall, ions such as NO_3^-, Cl^-, and SO_4^{2-} are considered to be "mobile" in soil. Conversely, ions such as Ca^{2+}, Mg^{2+}, K^+, Fe^{3+}, Mn^{2+}, Zn^{2+}, Cu^{2+}, and orthophosphate ($H_2PO_4^-$), which are mainly adsorbed onto soil particles, are "immobile."

Mycorrhizas and Nutrient Uptake

Mycorrhizas are a symbiotic association between a fungus and a plant root (the host). The fungus enhances the host plant's uptake of immobile nutrients such as P and, in return, obtains organic compounds for its growth. The most widespread type of mycorrhiza is an endomycorrhiza, so called because the fungus grows mainly inside the root where it develops branched structures called arbuscules that release nutrients inside the root cells (figure 3.4). This association is called an arbuscular mycorrhizal fungus (AMF).

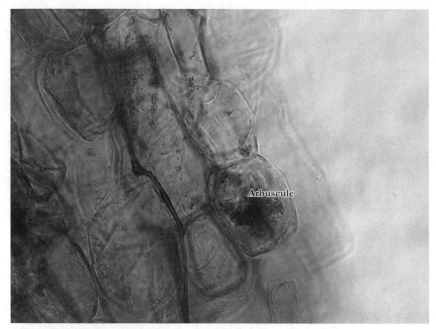

Figure 3.4 Example of an arbuscular mycorrhizal fungus infecting a grapevine root. The branched fungal structure inside one of the host cells is an arbuscule through which nutrients and carbon compounds are exchanged between the host and fungus. (Photo prepared by Dr. Melanie Weckert, National Wine & Grape Industry Centre, Wagga Wagga, Australia.) See color insert.

AMF are commonly found on *Vitis vinifera* roots and rootstocks. There are many strains of the fungus that survive as spores in the soil when no suitable host is present and have no preference for a particular host. However, in a perennial plant such as *V. vinifera*, less effective strains may come to predominate over time. Mid-row cover crops and weeds can counteract this trend by maintaining a more diverse AMF community in vineyard soils. Colonization by AMF is important for newly planted vines, especially in infertile soils.

The beneficial effect of AMF in enhancing P uptake is primarily through the fungal mycelium (see table 5.1, chapter 5), which provides a pathway for rapid P transport into the root, thereby short-circuiting the depletion zone shown in figure 3.3. This could also be the mechanism for enhanced uptake of other immobile elements such as Cu, Zn, and K that is sometimes observed. The mycorrhizal symbiosis is self-regulating in that, as soil P availability increases, the incidence of the infection declines. Mycorrhizal roots live longer than non-mycorrhizal ones and have an improved capacity to absorb water.

Nutrient Availability in the Soil

When considering their availability to grapevines, the nutrients and other elements can be divided into groups as follows:

1. Macronutrient elements N, S, and P that exist mainly in organic forms
2. Calcium, Mg, K, Na, and P, which bond with varying strength as inorganic ions to soil particles (note that P is included in both Groups 1 and 2)
3. Micronutrient elements Fe, Mn, Zn, Cu, and Mo, and trace elements Pb, Cd, Cr, and Ni, which form insoluble compounds and also bond strongly to mineral particles and organic matter
4. Chlorine and B (nutrients) and As (potentially toxic), which occur mainly in solution.

Only a nutrient in the soil solution is immediately available to vine roots. Assessing the availability of other nutrients that are distributed between solid and solution phases is difficult because both phases can contribute to the supply. Deficiency occurs when a nutrient's supply is insufficient to meet a vine's demand; conversely, when supply exceeds demand or a nonnutrient element accumulates in the root zone, toxicity may occur.

Availability of Nitrogen, Sulfur, and Organic Phosphorus

About 99% of soil N is in an organic form not directly available to a vine. Thus a soil with an organic C content of 1%, equivalent to 20,000 kg C/ha.15 cm,[2] will have between 1333 and 2000 kg N/ha.15 cm in an organic form, depending on the C-to-N ratio of the organic matter (discussed later). The available forms (NH_4^+ and NO_3^- ions) are produced through mineralization by microorganisms—literally, the conversion of an organic form into an inorganic or "mineral" form. Mineralization also describes the conversion of organic S and P into inorganic S (SO_4^{2-}) and P ($H_2PO_4^-$ and HPO_4^{2-}), which are the plant-available forms. The reverse process of incorporation of C, N, P, and S into living microbial tissue is called immobilization. The role of soil microorganisms in the C cycle is discussed in "The Soil Biomass," chapter 5. Box 3.2 describes some consequences of N mineralization and nitrification.

---

[2] This notation means kg C per ha to 15 cm depth.

Box 3.2 Mineralization, Nitrification, and Their Effects on Soil pH

The first step in mineralization of organic N—the production of NH_4^+ ions—is carried out by a range of heterotrophic microorganisms and is called ammonification. (Table 5.3 in chapter 5 explains the differences between heterotrophic and autotrophic organisms). Because OH^- ions are also produced to electrically balance the NH_4^+ ions, this step creates alkalinity.

Ammonium ions are taken up by vine roots, but much of the soil NH_4-N is consumed by specialist autotrophic bacteria that derive energy for growth by oxidizing NH_4^+ first to nitrite (NO_2^-) and then to nitrate (NO_3^-). This two-step process, carried out by various species of *Nitrosomonas* and *Nitrobacter* bacteria, respectively, is called nitrification and produces $2H^+$ ions for every NH_4^+ oxidized. The net outcome of ammonification followed by nitrification is that one mole each of NO_3^- and H^+ are produced for every mole of organic N that is oxidized (moles and moles of charge are explained in box 3.4).

Although nitrification is potentially acidifying, this potential is not realized if all the NO_3^- produced is taken up by plant roots, releasing HCO_3^- and OH^- ions in exchange, which cancel out the H^+ ions. However, NO_3^- is readily leached from the root zone by percolating water, in which case actual acidification takes place because the leached NO_3^- is accompanied by Ca^{2+} and other exchangeable cations that are displaced by the H^+ ions.

Soil acidification always occurs when an NH_4-N fertilizer such as ammonium sulfate (($NH_4)_2SO_4$) is added to soil, because when the NH_4^+ ions are oxidized, two H^+ ions are released for every NO_3^- ion produced, with no counterbalancing OH^- ions. The effect of this acidification is usually measured in terms of the "lime ($CaCO_3$) equivalent" required to neutralize the H^+ ions (i.e., to raise the soil pH). Experience shows that the practical lime equivalent of ($NH_4)_2SO_4$ is about 5.4 kg pure $CaCO_3$ per kg N in the fertilizer and about 1.8:1 for ammonium nitrate (NH_4NO_3) fertilizer.

The effect of urea fertilizer (($NH_2)_2CO$) on soil pH is similar to what occurs when soil organic N compounds are oxidized, with one important difference. The soil pH around a granule of urea initially increases to nine or more as ammonium hydroxide (NH_4OH) is formed from the urea. However, NH_4OH is unstable and breaks down to release NH_3 gas and H^+ ions, which neutralize the OH^- ions. The extent of any subsequent acidification then depends on how much of the remaining NH_4^+ is oxidized to NO_3^- and how much of this NO_3^- is leached from the soil.

The balance between mineralization and immobilization depends on whether there is enough of an element in an organic substrate to satisfy the microorganisms' demand for growth. The demand for N, S, and P is directly linked to the demand for C, the main building material of microbial cells. For example, if the C-to-N ratio in the substrate is small, there is likely to be surplus N that is released as NH_4^+, and thus net mineralization occurs. Conversely, if the C-to-N ratio is large, all the N is incorporated into microbial tissue and some mineral N may also be taken from the surrounding soil—net immobilization then occurs. Figure 3.5 shows that the critical C-to-N ratio for the tipping point between net mineralization and net immobilization is about 25.

As table 3.2 shows, materials such as cereal straw or vine prunings, which have high C-to-N ratios, are likely to induce net N immobilization in the soil. Low C-to-N ratio materials such as legume residues from a mid-row cover crop, for example, should promote net N mineralization. Although net mineralization can occur when well-humified organic matter decomposes, because its decomposition is slow, the release of mineral N is small compared with that released from fresh legume residues. However, some naturally fertile soils with more than 5% organic C and a low C-to-N ratio can mineralize substantial amounts of N that may predispose to too much N uptake and excess vigor in grapevines.

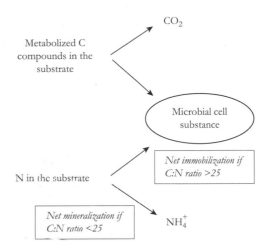

Figure 3.5 How the carbon-to-nitrogen ratio affects the balance between net mineralization and immobilization of nitrogen in the soil. (Redrawn from White, 2003.)

Table 3.2 **Expected Net Nitrogen Mineralization from Soil Organic Matter or Returned Plant Residues**

Organic material	C-to-N ratio	Net mineralization expected in one year (kg N/ha)
Well-humified soil organic matter	10–15	Small (10–15)
Legume residues (clover, medics, peas, beans)	15–25	Significant (25–50)
Grape leaves	30	<5
Cereal straw	40–120	Nil
Winter prunings	100	Nil

Similar principles apply to the mineralization of organic P and S compounds in soil, except that the critical C-to-P and C-to-S ratios are of the order of 10 times higher than the C-to-N ratio.

Availability of Calcium, Magnesium, Potassium, Sodium, and Inorganic Phosphorus

The available forms of Ca, Mg, K, and Na are the cations Ca^{2+}, Mg^{2+}, K^+, and Na^+. These cations are attracted to negatively charged soil particles and can be exchanged by other cations in the soil solution; hence they are called exchangeable cations. Although not a nutrient, Na^+ is included because Na^+ ions have an important effect on the behavior of soil colloidal particles in water (see box 4.3, chapter 4). In addition to its organic component, half or more of the P in soil may occur in an inorganic form that is strongly bonded to Fe and Al oxides or precipitated as an insoluble calcium phosphate.

Within a soil's clay fraction, identified in "Calibration for Texture," chapter 2, the negatively charged particles comprise crystalline clay minerals and organic matter. Iron and Al oxides are normally present as positively charged particles or as "coatings" on clay surfaces. All these components of the clay fraction are important for nutrient retention in soil, as discussed in the next section.

Retention of Nutrients by Clay Minerals and Oxides

Although clay minerals and oxides are of varied composition, they exhibit several basic features, outlined in box 3.3. Examples of the crystalline structure of the common clay minerals are shown on the website www.virtual-museum. soils.wisc.edu/displays.html. Because minerals of the clay fraction are very small (<2 microns equivalent diameter), they have a large surface area per unit volume (called the specific surface area). The mineral montmorillonite, which belongs to the same group as the bentonite used to coagulate wine proteins, has the largest specific surface area. For example, if a heaped teaspoon (about 5 g) were spread out one layer thick, the clay particles would cover about half a football field.

The large specific surface area and permanent negative charge make clay minerals the site of many chemical and physical reactions in soil. The negative charge is measured by the number of moles of cation charge that can be held per kilogram of clay. This is called the cation exchange capacity (*CEC*). However, because organic particles can also carry a negative charge (see next section), *CEC* is usually measured as the moles of cation charge per kg of soil.

Box 3.4 explains the concepts of moles and moles of charge, and table 3.3 summarizes some of the properties of the common clay minerals.

Box 3.3 Some Basic Features of Soil Clay Minerals and Oxides

Clay minerals have a plate-like crystalline structure for which the flat surface area is large relative to the edge face area. Figure B3.3.1 shows the basic form of a clay crystal. There is some substitution of ions of similar size but different positive charge in the crystal structure, which results in a permanent negative charge in the particle. The most common substitutions are Al^{3+} for Si^{4+}, Mg^{2+} for Al^{3+}, and Fe^{2+} for Al^{3+}.

Cations in the soil solution are attracted by the permanent negative charge to the flat clay surfaces—these cations are called exchangeable cations. Depending on the type of clay mineral, the much smaller edge faces can also attract cations when the solution pH is above 6 (the edge faces are negatively charged then). At pH <6, the edge faces are positively charged and hence attract anions.

Like the clay mineral edges, Fe and Al oxides show a variable charge that changes from positive to negative as the solution pH increases—this is referred to as pH-dependent charge. Up to about pH 8, the oxides are predominantly positively charged and thus are the main sites for the adsorption of anions such as $H_2PO_4^-$, Cl^-, NO_3^-, SO_4^{2-}, and some polyanionic organic compounds. Of these, $H_2PO_4^-$ ions are the most strongly adsorbed because P has a strong affinity for Fe and Al and bonds chemically to these atoms at the mineral surface.

Highly weathered soils such as Kraznozems (red loams and clay loams rich in Fe and Al oxides), soils with high organic matter content, and soils derived from volcanic ash and tephra can be strongly influenced by pH-dependent charges. Understanding these phenomena is important for the correct management of the chemical and physical condition of soils.

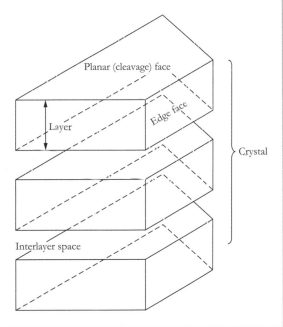

Figure B3.3.1 Diagram of the structure of a clay crystal showing crystal layers, planar (flat) surfaces, and edge faces. Such a crystal cannot be seen without a microscope.

Box 3.4 Moles and Moles of Charge in Soil Chemistry

A mole of an element or compound is its molecular weight in grams, more commonly referred to as the molar mass. The standard against which all substances are measured is the stable isotope of carbon (^{12}C). On this scale, the molar mass of hydrogen (H) is 1 g, calcium (Ca) is 40 g, and aluminum (Al) is 27 g.

The recommended unit of charged mass for cations, anions, and charged surfaces is the mole of charge, which is equal to the molar mass divided by the ionic charge. It follows that

- For H^+, because its ionic charge is +1, a mole of charge is 1/1 = 1 g.
- For Ca^{2+}, because its ionic charge is +2, a mole of charge is 40/2 = 20 g.
- For Al^{3+}, because its ionic charge is +3, a mole of charge is 27/3 = 9 g.

For clay minerals, oxides, and organic matter, the most appropriate unit is the centimole of charge (+) or (−) per kilogram (abbreviated cmol (+)/kg). For example, the *CEC* of clay is expressed in cmols (+) per kg because *CEC* is measured by the moles of cation charge adsorbed by the clay. The *CEC* expressed in this way is the same as that expressed in the obsolete units, still used in some laboratories, of milliequivalents (mEq) per 100 g.

Table 3.3 Some Features of the Common Clay Minerals

Clay mineral type	*CEC* (cmol (+)/kg)	Specific surface area (m²/g)	Other salient features
Kaolinite	5–25	5–40	Strong bonding between crystal layers; interlayer spaces not accessed by cations or water; relatively large crystals and minimal swelling
Illite	20–40	100–200	K^+ ions fit snugly into holes between opposing crystal surfaces and hold the layers together; as K^+ is lost by weathering and replaced by partially hydrated Ca^{2+} and Mg^{2+}, the crystal expands slightly
Vermiculite	150–160	300–500	This is the end product of illite weathering in which K^+ ions have been replaced by Ca^{2+} and Mg^{2+}; limited swelling
Montmorillonite	100–120	750[a]	Crystal layers are held in roughly parallel alignment when Ca^{2+} and Mg^{2+} are present; swelling increases as these ions are replaced by Na^+ and dispersion can occur (see box 4.3, chapter 4)

Note. CEC = cation exchange capacity.

[a] For montmorillonite saturated with Na^+ ions.

Retention of Nutrients by Soil Organic Matter

Soil organic matter has a heterogeneous composition not only because of the variety of organic residues deposited on the soil but also because of the many biochemical processes, driven by microorganisms, which occur in soil. The role of polysaccharides and polyuronide gums (sticky materials) in stabilizing soil structure is discussed in "Soil Aggregation," chapter 4. Here I describe the properties of charged compounds, such as polymerized phenols and organic acids, which are important constituents of humified organic matter.

As plant and animal residues decompose, a dark, humic material slowly forms, most easily seen beneath the litter from broadleaf trees or grasses (figure 3.6), or in well-made compost (see "Organic Viticulture"). Humification leads to an increase in the number of organic chemical groups that can dissociate H^+ ions and become negatively charged. These groups are weak acids of varying strength so that they gradually release H^+ as the pH increases from less than 3 to more than 8.

The negative charge on the organic matter increases as the pH increases and this translates into an increase in CEC. Compared on a unit weight basis, humified organic matter has 3 to 10 times the CEC of illite or montmorillonite clay

Figure 3.6 Example of well-humified organic matter formed at the soil surface under grass litter.

at pH 7, so it makes a significant contribution to the total *CEC* of any soil that is sandy and has more than 1% to 2% organic C. In clay soils, the contribution of organic matter relative to that of the clay minerals is much less because of the preponderance of clay relative to organic matter.

The variable negative charge on organic matter attracts cations such as Fe^{3+}, Al^{3+}, and Ca^{2+}. These adsorbed cations can act as an "electrostatic" bridge between the organic matter and negatively charged clay minerals. Alternatively, negatively charged organic polymers can be directly attracted to positively charged Fe and Al oxides. These kinds of charge interactions are fundamental to the aggregation of soil particles, which is a prerequisite for good soil structure.

Availability of the Micronutrients and Trace Elements

Iron, Mn, Cu, and Zn occur as metallic cations (see table 3.1), as do the trace metals Cd, Pb, and Ni (as Cd^{2+}, Pb^{2+}, and Ni^{2+}). The fully charged cationic forms exist only at low pH. As the pH increases, these hydrated metal cations hydrolyze[3] and eventually precipitate as insoluble hydroxides. For example, the cation $Fe(H_2O)_6^{3+}$, which exists at pH less than 3, hydrolyzes to the hydroxy form $Fe(OH)(H_2O)_5^{2+}$ as the pH increases. This hydrolysis continues with the stepwise release of H^+ ions until neutral $Fe(OH)_3$ is formed between pH 4 and 5. The other micronutrient cations and trace element metals (Pb, Ni, and Cd) behave similarly, with their hydroxides forming at higher pHs. Because of adsorption of the hydroxyl cations by clays and precipitation of the hydroxides, the overall effect of an increase in pH is for the availability of these elements to decrease.

The metal cations Fe^{3+}, Mn^{2+}, Cu^{2+}, and Zn^{2+} also form complexes with organic compounds. Although these complexes are generally insoluble (when the organic matter is well humified), repeated applications of the metals can raise the concentration of the soluble fraction of the complexes to a point where they become toxic to beneficial soil organisms such as earthworms. This effect has been observed in older vineyards that have been repeatedly sprayed with Bordeaux mixture (a mixture of lime in copper sulfate [$CuSO_4$] solution). On the other hand, some synthetic organic compounds called chelates are used to form soluble complexes with Fe, Mn, Cu, and Zn to inhibit the formation of the insoluble hydroxides and make these nutrients more plant-available. One example is the synthetic compound ethylenediaminetetraacetic acid (EDTA), which forms the complex Fe EDTA$^-$, which is more stable than $Fe(OH)_3$ up to pH 9. Similar chelation occurs between metal cations and natural organic

---

[3] Hydrolysis occurs when a water molecule attached to a cation splits into H^+ and $OH-$ and the H^+ ion is released.

compounds to form complexes that can be used as micronutrient fertilizers (see "Micronutrient Fertilizers").

Molybdenum occurs as the molybdate anion (MoO_4^{2-}), which behaves like $H_2PO_4^{-}$. It is strongly adsorbed to Fe and Al oxides at low pH and becomes more available as the pH increases. Boron occurs as neutral boric acid (H_3BO_3) up to pH 8, when it begins to dissociate and release the borate anion $B(OH)_4^{-}$, which is weakly adsorbed by oxides.

Testing for Deficiencies and Toxicities in Grapevines

What You Can See

The first sign of a nutritional problem may be a "visual symptom"—an abnormality in growth, leaf color, flowering, or fruit set—that suggests one or more of the essential elements is in short supply, or perhaps there is a toxicity, as might occur with too much Na or Cl. Where the symptoms appear depends on an element's mobility. An element is mobile if it exists as an ion in the tissue (e.g., K^+, Na^+, and Cl^-) or because it is in high demand and is translocated from older, mature parts to young growing tissues (e.g., N and P). Some of the more obvious visual symptoms are

- Chlorosis, referring to leaf yellowing resulting from a lack of chlorophyll. Nitrogen deficiency typically causes overall chlorosis, whereas Fe, Mg, and Zn deficiencies cause yellowing between the leaf veins (interveinal chlorosis)
- Necrosis, referring to the death of leaf tissue. As a deficiency becomes more severe, leaves may discolor and begin to die. Necrosis can also start as a scorching of leaf margins, as in the case of Cl toxicity
- "Hen and chickens" appearance, which refers to bunches having a few normal berries and many small, seedless berries. This can occur with Zn and B deficiencies.

Table 3.4 summarizes nutrient deficiency symptoms, and figure 3.7 shows examples of deficiencies. Goldspink and Howes (2001) give further details of deficiency and toxicity symptoms in grapevines.

Plant Analysis

Diagnosis of deficiencies and toxicities can be confusing when symptoms caused by different elements are similar (e.g., Fe, Mg, and Zn). Also, visual symptoms appear only after the plant has suffered a check to growth as a result

(a)

(b)

Figure 3.7 (A) Iron deficiency in a vineyard in the St. Emilion appellation, Bordeaux region, France. Note yellowing of the younger leaves and marked interveinal chlorosis. (B) Zinc deficiency in grapevines on a high pH soil in the Riverina region, New South Wales, Australia. Leaves are not as severely chlorotic as with Fe deficiency in figure 3.7A. See color insert.

Table 3.4 **Nutrient Mobility and Visual Symptoms of Deficiencies in Grapevines**

Nutrient element	Mobility and appearance of symptoms
N	Mobile; overall reduction in growth and uniform leaf yellowing
P	Mobile; bronze coloration in older leaves early in the season
K	Mobile; starts as a yellowing of older leaf margins; as deficiency worsens, margins die and curl upwards
S	Mobile; deficiency rare in vineyards where S sprays are used to control powdery mildew; symptoms similar to N deficiency
Ca	Very immobile; shoot tips stunted and may die
Cl	Mobile; no deficiency seen because Cl is ubiquitous and readily absorbed
Fe	Immobile; young leaves show interveinal chlorosis; when severe, leaves may become very pale before developing necrotic (dead) spots
Mg, Mn	Mobile; interveinal chlorosis starts on older leaves and, when severe, necrosis extends inward from the leaf margins
Zn	Relatively immobile; stunted lateral shoots with small leaves showing "blotchy" interveinal chlorosis; fruit set may be affected; bunches show a "hen and chickens" appearance
Cu	Immobile; shortened internodes and death of shoot tips similar to Ca; flowering also affected; rare in vineyards because of Cu sprays
B	Immobile; occurs in older leaves and may be confused with Zn deficiency because of "hen and chickens" appearance; the range between deficiency and toxicity is small
Mo	Immobile; necrosis of leaf margins; affects pollen tube growth and fertilization, resulting in poor fruit set

of a "hidden hunger" for the element, as illustrated by the relationship between fruit yield and a nutrient's concentration in vine leaves (figure 3.8). This relationship is the basis for diagnosing a vine's nutrient status by plant analysis or tissue testing. Figure 3.8 shows that as yield increases up to a maximum with increasing nutrient supply, the nutrient concentration in the tissue also increases. The degree of deficiency is indicated by where the plant's analysis lies on the approximately linear trendline between severe deficiency and the optimum range.

Figure 3.8 also shows the critical value of nutrient concentration below which an increase in supply leads to increased yield. Note that in the "luxury consumption" range, yield may decline as a result of nutrient imbalances or outright toxicity. For example, too much N encourages vigorous growth that can cause excessive shading of basal buds, which leads to lower fruitfulness the following season. Excess N can predispose to "false potassium deficiency" under cool and variable weather in spring due to a K/N imbalance and may also cause early bunch stem necrosis.

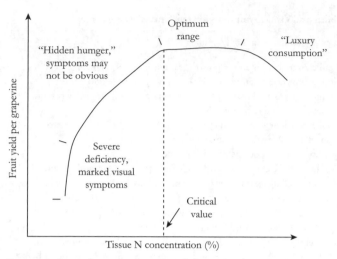

Figure 3.8 Stages from severe nutrient deficiency to luxury consumption according to tissue concentrations. (White, 2003)

Sampling Protocol for Plant Analysis

Which Part?

Because nutrients can be remobilized in response to a change in the site of greatest demand as a vine grows, the part to be sampled for analysis must be specified. It is generally recommended that leaf petioles be sampled because they are more responsive to a vine's nutrient status than leaf blades. Whether blades or petioles are sampled, washing and damp-drying are advisable to avoid the effect of dust on the analysis.

The leaf sampled should be a basal leaf opposite a bunch—usually the fourth or fifth leaf from the base of a shoot, as shown in figure 3.9A. The petiole is separated from the blade and placed in a paper bag, to be sent to a laboratory for analysis as soon as possible. If toxicity is suspected, leaf blades may also be sent for analysis because elements such as B accumulate more in the blade than the petiole. If blades are sampled, the most recently matured leaf on a shoot is recommended—usually the fifth to seventh leaf behind the tip (figure 3.9B).

Time of Sampling

The concentration of nutrients such as N, P, and K is greatest in young leaves and declines as the leaf ages. Table 3.5 gives an example of how the concentrations of mobile elements can change with the stage of growth. Traditionally, petioles are

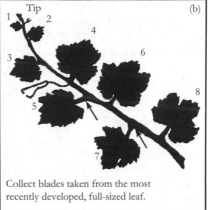

Select petiole opposite either of the basal clusters during full bloom.

Collect blades taken from the most recently developed, full-sized leaf.

Figure 3.9 (A) Position of leaves to be sampled for tissue analysis (petioles) at flowering (bloom). (B) Sampling blades from mature leaves. (Christensen et al., 1978; reprinted with permission of the Division of Agriculture and Natural Resources, University of California.)

Table 3.5 An Example of Changes in the Concentration of Some Mobile Nutrients in Petioles over Time

	Date of sampling[a]		
	November 9, 2012 (End of flowering)	November 14, 2012 (Fruit set)	February 7, 2013 (Postveraison)
N concentration (%)	2.2	2.0	0.54
P concentration (%)	0.86	0.69	0.62
K concentration (%)	1.9	1.2	0.53
S concentration (%)	0.26	0.24	0.15

[a] For Shiraz vines in a vineyard in the Yarra Valley region, Victoria, Australia.

sampled at full flowering or bloom when the leaf is more sensitive to the external nutrient supply than at later stages (box 3.5). The argument for even earlier sampling is that a grower has more time to respond with a remedial fertilizer treatment, but the disadvantage is that concentrations are changing more rapidly then. The exact time of sampling is not so critical if a regular sampling program (every year or second year) is followed (as discussed in "Interpreting Plant Analyses").

For troubleshooting, or trying to confirm a deficiency or toxicity indicated by visual symptoms, leaves may be collected at any time. Leaves are best sampled early in the day when they are least likely to suffer water stress and must not be sampled soon after a foliar spray has been applied.

Box 3.5 Stages in Grapevine Growth and Fruit Ripening

Bud burst (bud break in Europe)—bud dormancy is broken and young leaves begin to emerge and expand in spring (early October or April in cool-climate regions of the southern and northern hemispheres, respectively). *Flowering* (bloom in North America)—may extend over several weeks in cool climates but usually 10 to 14 days. The midpoint occurs when 50% of fruit clusters (bunches) are in flower; cell division begins in the fertilized flowers. *Fruit set*—may extend over a few weeks, but the midpoint occurs when 50% of bunches have berries 3 to 5 mm in diameter; all unfertilized berries should have been shed; from fruit set to veraison, berry growth is by cell division and expansion. *Veraison*—berries begin to soften and change color (white varieties begin to turn yellow and red varieties turn red) as ripening begins and sugars start to accumulate. Midveraison occurs when 50% of bunches are coloring (for red grapes); after veraison (postveraison), berry growth is mainly by cell expansion. *Maturity*—full ripeness attained; red grapes fully colored, stems turn woody, bunches are harvested. *Postharvest*—vines continue to grow, especially through root growth, provided temperature and soil water are adequate; leaves turn yellow and die.

Selecting the Sample Material

The traditional method is to take a composite sample from a block of vines of the same variety on each soil type. Seventy-five to 100 petioles are collected from randomly chosen vines in the block. Separate composite samples should be taken from areas that are performing badly or well.

Even on one soil type, measurements at a very high density may show considerable variation in the nutrient concentrations over short distances. Figure 3.10 shows the variation in petiole P concentration in a Cabernet Sauvignon vineyard in the Coonawarra region, South Australia. Such variation probably reflects the influence of soil physical properties on growth, in addition to an effect of the soil's P supply. Thus a strong case can be made for targeted sampling based on the known pattern of soil variability, which goes further than the recommendation that separate samples be taken from "good" and "bad" areas. However, the cost of the extra analyses must be balanced against the potential value added by the more precise information, so methods of analyzing many samples rapidly and cheaply need to be considered, as discussed in this chapter in "Precision Viticulture for Better Soil Management."

Interpreting Plant Analyses

Analyses for micronutrients are reported in milligrams per kilogram dry matter and as percentage dry matter for the macronutrients (details are given in box 3.1). The result always has some uncertainty because of natural variability in the soil

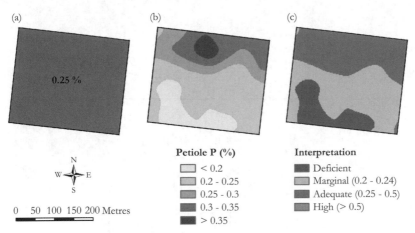

(a)   (b)   (c)

0.25 %

N
W ← → E
S

0   50 100 150 200 Metres

Petiole P (%)
☐ < 0.2
☐ 0.2 - 0.25
☐ 0.25 - 0.3
■ 0.3 - 0.35
■ > 0.35

Interpretation
■ Deficient
☐ Marginal (0.2 - 0.24)
■ Adequate (0.25 - 0.5)
☐ High (> 0.5)

Fig. 3.10 (A–C) Maps of petiole phosphorus (P) concentration in a small Cabernet Sauvignon vineyard in the Coonawarra region, South Australia, assessed by the "industry standard" approach of analyzing a bulk sample (A) and intensive sampling of vines on a grid pattern (B). Although the same criteria are used to interpret maps A and B, the map (C) that is based on (B) presents a different conclusion with respect to P sufficiency from A. (Courtesy of Dr. Robert Bramley, CSIRO Ecosystem Sciences, Adelaide, South Australia.)

and vines and because of analytical error. This uncertainty can be minimized with the best sampling and laboratory procedures, but too much emphasis should not be placed on a single set of results. Variations in weather from season to season will affect leaf nutrient concentrations, so the best practice is to gather test results over several seasons to show trends in nutrient status. As shown in figure 3.11, any anomalous result will then become obvious.

Record the condition of the vines and soil (especially soil moisture) at sampling time, as well as recent fertilizer or manure applications and cultural operations such as spraying with a fungicide. This information, together with the results of plant analyses and any soil tests, provides a valuable record when compiled for individual blocks over several seasons and should be part of any integrated production system, as discussed in "Integrated Production Systems and Sustainability," chapter 6.

Using Critical Values

A critical value for each nutrient is ideally determined for an individual variety when growth is not limited by any other nutrient or the water supply (critical values for plants under water stress tend to be lower than for well-watered plants). However, varieties whether on own roots or rootstocks may differ in their nutrient uptake and yield–nutrient relationships. This makes the task of calibrating plant analyses for commercial combinations of variety and rootstock so large that only a general set of values can be given, based on the much-studied Sultana variety

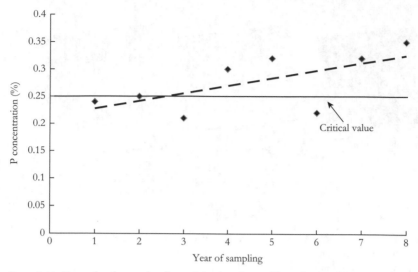

Figure 3.11 Example of a crop log for petiole phosphorus (P) analysis. Values in years 3 and 6 less than the critical value can be ignored because the general trend is upward.

Table 3.6 Macronutrient Concentrations in Petioles for Assessing the Nutrient Status of Grapevines

Macronutrient element	Very deficient (%)[a]	Deficient (%)[a]	Low to marginal (%)[a]	Adequate (%)[a]	High to excessive (%)[a]
N (total)		<0.7	0.7–0.89	0.9–1.2	>1.2
N as NO_3^- (mg/kg)		<340	340–599	600–1500	>1500
P	<0.15	0.15–0.19	0.2–0.24	0.25–0.5	>0.5
P (Pinot Noir only)	<0.12	0.12–0.14	0.15–0.19	0.2–0.4	>0.4
K (with adequate N)		<1.0	1.0–1.7	1.8–3.0	>3.0
Ca		<1.2		1.2–2.5	
Mg		<0.3	0.3–0.4	>0.4	
Na					>0.5
Cl					>1.0

[a] Except where indicated otherwise.

Compiled from Robinson et al. (1997), Goldspink and Howes (2001), and Robinson (2005).

(Thompson seedless) in the United States. These general values have been adapted to Australian conditions using data from vineyards in South and Western Australia. Tables 3.6 and 3.7 give petiole concentration ranges for macronutrients and micronutrients, respectively. The critical values occur between the "marginal" and "adequate" categories.

Table 3.7 **Micronutrient Concentrations in Petioles for Assessing the Nutrient Status of Grapevines**

Micronutrient element	Deficient (mg/ kg)	Low to marginal (mg/kg)	Adequate (mg/kg)	High to excessive (mg/kg)
Fe		25–30	>30	
Cu	<3	3–5	6–11	>40[a]
Zn	<15	15–25	26–150	>450[a]
Mn	<20	20–29	30–60	>500[a]
B	<25	25–34	35–70	>100
Mo	<0.05	0.05–0.19	0.20–99	>100

[a] May indicate spray contamination.

Compiled from Robinson et al. (1997), Goldspink and Howes (2001), and Robinson (2005).

Assessment of a vine's N status is improved when NO_3–N concentration is used together with a total N analysis, but general leaf color and vine vigor should also be considered. There is an interaction between N and K in vines that results in petiole K concentrations being lower when vines are well supplied with N. For this reason, K analyses should always be interpreted in conjunction with N. Ideally, the K-to-N ratio should be between 1 and 2.

Soil Testing

Because it is a perennial, the grapevine is best suited to plant analysis, when repeated samples can be collected from the same plant from year to year. Notwithstanding that soil testing is more widely used for annual crops, soil analysis still has value for grapevines, especially at vineyard establishment when it is the only tool available (see "Describing a Soil Profile," chapter 2). A soil test is essential for pH, from which a decision is made to apply lime if the soil pH is too low (see box 3.6) or whether a micronutrient deficiency might occur if it is too high. In France, a soil test is used to estimate "active" calcium carbonate, which if too high can reduce Fe availability; soil tests are also valuable for indicating salinity and sodicity (see "Soil Testing for Salinity").

Various tests have been developed to measure key chemical, physical, and biological properties of soil. Here I deal briefly with the chemical tests; relevant physical tests are discussed in chapter 4 and biological tests in chapter 5. The prime purpose of a chemical test is to assess the soil's ability to supply an essential nutrient or to determine whether there is a potential toxicity, as in the case of salinity for example. Although analytical methods have been developed to extract the "available" part of the total element in a soil, this is but a snapshot in time. Moreover, this concept of availability is incomplete, because a vine's demand for a

Box 3.6 Testing for Soil pH and Lime Requirement

Soil pH indicates the likely effect of H^+ ions on plant growth and the potential for toxicities and deficiencies to occur. The pH stands for the negative logarithm of the H^+ ion concentration, measured in moles per liter (M) of solution. The scale is from 0 to 14 pH units, with pH 7 being the neutral point (equal concentrations of H^+ and OH^- ions).

Soil pH is measured in the field with a universal indicator and color chart to an accuracy of ±0.5 pH units; alternatively, in a laboratory it is measured with a pH meter to an accuracy of ±0.05 pH units. Because soil is mixed with water (usually one part by weight of soil to five parts by volume of distilled water), the laboratory pH differs from the field pH. To minimize this difference, the soil can be shaken with a solution of 0.01 M $CaCl_2$ (1:5 ratio), which gives a pH value about 0.5 to 0.8 units lower than in water and closer to the field pH. For this reason, it is important to know which method of measurement has been used when pH values are quoted.

The recommended $pH(CaCl_2)$ for grapevines is between 5.5 and 7.5. Vines will grow outside this range but are more likely to suffer a toxicity or deficiency. For example, the availability of Fe, Cu, Zn, and Mn decreases as the soil pH increases; conversely, at a $pH(CaCl_2)$ of less than 5.5, the hydrated aluminum cation $Al(H_2O)_6^{3+}$ begins to accumulate and can impair root growth and P uptake. In fact, clay in very acidic soils ($pH(CaCl_2)$ <4.5) is dominated by these Al^{3+} ions. Hydrolysis of the hydrated Al^{3+} ions produces a continuing supply of H^+ ions, which means that considerable amounts of lime are required to raise the pH to an acceptable value of 5.5 or above. A special laboratory soil test involving titration of a soil with an alkali is used to measure the lime requirement, which can range from 1 to 5 t $CaCO_3$/ha.

Laboratory analysis is also necessary to measure the neutralizing value (NV) of a liming material. Table B3.6.1 gives the NV of several such materials. Particle size is also important, because the smaller the particles, the faster they dissolve in the soil. A high-quality liming material should have an NV greater than 85% and more than 60% of the particles less than 0.15 mm in diameter. Because lime is slow to dissolve, it is most effective when cultivated into the soil so that the contact between soil and lime particles is maximized.

Table B3.6.1 Liming Materials for Vineyards

Material and its chemical composition	Neutralizing value of commercial grade (%)[a]	Comments
Burnt lime, CaO	>150	Reacts vigorously with water
Hydrated lime, $Ca(OH)_2$	120–135	Occurs as a very fine powder; difficult to handle
Dolomite, $(Ca,Mg)CO_3$	95–110	More soluble than $CaCO_3$; contains about 11% Mg
Limestone, $CaCO_3$ plus impurities	50–85	Neutralizing value depends on the concentration of impurities such as clay, silica, and Fe and Al oxides
Cement kiln dust, $CaCO_3$, CaO, K	90–110	Fine powder, by-product of cement manufacture; often pelleted; also supplies K

[a] Calculated relative to pure $CaCO_3$ as 100%.

Compiled from Goldspink and Howes (2001) and the Fertilizer Federation Industry of Australia (2006).

nutrient varies with its genetic potential, and many of the soil processes affecting a nutrient's supply to roots are dynamic.

Apart from pH measurement, the most common soil tests are for available P, S, and the exchangeable cations Ca^{2+}, Mg^{2+}, K^+, and Na^+. Soil testing for N is less common, mainly because the amount of available N (NH_4^+ plus NO_3^-) is biologically dynamic (see box 3.2). Because many chemical extractants have been used, especially for P, the analytical method must be known in interpreting a soil test result. Not only do analytical methods vary for any one nutrient, but different laboratories can also produce different results for the same test on samples of the same soil. Table 3.8 gives an example of interlaboratory variation in soil test results for soil from a single vineyard. Clearly, field sampling error and laboratory analytical error can engender uncertainty in a set of soil test results, especially for nutrients such as available N and P. However, by following a consistent sampling protocol, using the same commercial laboratory, and making measurements over time, as recommended for plant analysis (see figure 3.11), a winegrower can see how a soil test value is trending and minimize uncertainties introduced by temporal, spatial, and human factors.

In addition to differences in analytical methods, there are differences among commercial laboratories in the interpretation of soil tests. Box 3.7 discusses the concept of the basic cation saturation ratio (BCSR), commonly known as the Albrecht System, which is an example of an "alternative" interpretation of exchangeable cation values.

Table 3.8 **Soil Test Results from Two Laboratories for Samples (0–10 cm) of the Same Soil from a Vineyard in the Gippsland Region, Victoria, Australia**

	Laboratory A		Laboratory B	
Soil test and units	Under vine	Mid-row	Under vine	Mid-row
pH (water)	6.5	6.7	6.3	6.5
pH (CaCl$_2$)	5.8	6.0	5.8	6.0
EC (1:5) (dS/m)	0.12	0.12	0.15	0.16
Organic C (%)	2.6	4.1	2.3	3.6
Mineral N (NH$_4^+$ and NO$_3^-$) (mg/kg)	10	14	7	7
Bicarbonate-extractable P (mg/kg)	33	24	78	87
Available K (mg/kg)	273	351	281	320
Exchangeable Ca (cmol (+)/kg)	7.2	7.8	6.9	7.4
Exchangeable Mg (cmol (+)/kg)	1.3	1.3	1.2	1.4
Exchangeable Na (cmol (+)/kg)	0.2	0.1	0.4	0.4
CEC (cmol (+)/kg)	9.4	10	12	12

Box 3.7 The Albrecht System of a Basic Cation Saturation Ratio

The concept of a basic cation saturation ratio (BCSR) postulates that the exchangeable cations Ca^{2+}, Mg^{2+}, K^+, and Na^+ should be present in "ideal" proportions of a soil's cation exchange capacity to promote optimum plant growth. After research on a range of crops in the United States, Albrecht (1975) concluded that the ideal proportions were Ca 60% to 75%, Mg 10% to 20%, K 2% to 5%, Na 0.5% to 5%, and H 10%, and specific values were placed on the ratios of Ca-to-Mg, Ca-to-K, and K-to-Mg for optimum plant nutrition.

What came to be known as the Albrecht System of a balanced BCSR has been adopted in various countries. In Australia, for example, a commercial laboratory in Victoria offers a comprehensive soil testing service based on the Mikhail System of soil balance, which has similarities to the Albrecht System. Such has been the interest in the Albrecht System that two scientists at the University of Queensland undertook an extensive review of the concept of an "ideal" or "balanced" BCSR (Kopittke and Menzies, 2007). They concluded that within the range of values commonly found in soils, the proportions of Ca, Mg, and K ions and their specific ratios do not influence the chemical, physical, or biological fertility of a soil. Notwithstanding this conclusion, it is recognized, particularly in Australia, that a soil's exchangeable Na percentage is critical in determining a soil's structural stability (see box 2.4, chapter 2).

Mainstream soil science does not accept the concepts of an ideal BCSR and ideal cation ratios but focuses on assessing whether the concentrations of available nutrients are sufficient but not excessive. This approach is called the sufficiency level of available nutrients, which incorporates the concept of a critical value for an essential nutrient as determined through soil or plant analysis (see figure 3.8). This approach is potentially more conservative of fertilizer and soil amendment use than the BCSR approach and should therefore make better use of scarce resources.

Critical Values and Soil Sampling

The standard approach to interpreting a soil test value involves determining the range of test values that correspond to a very deficient, deficient, adequate, and excessive nutrient supply—the sufficiency level of available nutrients (see box 3.7). Although this information should ideally be derived from research trials in vineyards, much of these data have been sourced from the more abundant agronomic trials. A critical value is derived from a response curve, analogous to that shown in figure 3.8, except that soil test values are plotted on the horizontal axis. This approach is necessary to ensure soil test values are correctly interpreted, preferably by a specialist adviser. Also, the normal sampling depth of 0 to 10 cm or 0 to 15 cm may be inadequate for deep-rooted vines. In Western Australian vineyards, for example, it is recommended that soil samples be collected in 10-cm intervals down to 50 cm. This procedure is labor-intensive if done by augering (see figure 1.13A); a suitable alternative could be a hydraulic corer (see figure 1.13B). When sampling especially for salinity and sodicity, an appropriate compromise is to take samples from the surface soil (0–15 cm) and subsoil (45–55 cm). Chapter 6 deals with the topic of benchmarking soil health in vineyards on the basis of soil chemical, physical, and biological properties.

Soil Testing for Salinity

Soil salinity is routinely measured using a conductivity meter that responds to the total concentration of dissolved salts in the soil (see "Water Quality," chapter 2). A sieved sample of soil is made into a glistening paste with distilled water; the *EC* of the resultant solution—the saturation extract—is called the EC_e. In Australia, *EC* is commonly measured in a 1:5 soil-to-water extract so when soil *EC* values are reported, the method of measurement must be specified. Appendix 2 gives a table for converting from $EC(1:5)$ to EC_e.

Chapter 2 discusses soil surveys and making EM measurements, which can be calibrated to indicate salinity levels in salt-affected soils by means of a set of soil samples and a conductivity meter. Chapter 5 discusses the tolerance of *V. vinifera* and rootstocks to salinity. The effect of Na^+ ions in creating soil sodicity is discussed in box 2.4, chapter 2.

Correcting Nutrient Deficiencies and Toxicities

Among vines, soil, and the atmosphere, there is a natural cycling of nutrients in which the winegrower can intervene at various points. This intervention can be through foliar sprays (appropriate for correcting micronutrient deficiencies), fertilizers (mainly for the macronutrients N, P, K, Ca, and Mg), manures, and compost. Each of these methods is discussed in subsequent sections. The efficacy of nutrient management by these means depends on the point in the cycle at which intervention occurs and its timing in relation to the growing season.

Nutrient cycling is best illustrated with reference to N, which is a key nutrient affecting canopy growth, fruit yield, must fermentation, and ultimately wine quality.

Nitrogen Cycling

Figure 3.12 shows the N cycle in a vineyard. Atmospheric N_2 gas enters the cycle through legume N_2 fixation (see box 5.2, chapter 5). There are also "free-living" bacteria and Cyanobacteria in soil that can fix N_2, but their contribution in vineyards is small compared with that of legumes in cover crops.

The N-containing gases nitrous oxide (N_2O) and nitric oxide (NO) enter the atmosphere from natural or anthropogenic sources. Nitric oxide is oxidized in sunlight to nitrogen peroxide (NO_2), which dissolves in water droplets to

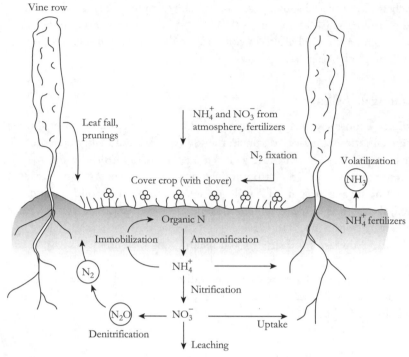

Figure 3.12 The nitrogen (N) cycle in a vineyard. (White, 2003)

form nitric acid and contributes to "acid rain." Ammonia gas released to the air from NH_4-N fertilizers, animal dung, and urine (see box 3.2) also dissolves in water droplets or is adsorbed as NH_4^+ ions on dust particles, to be returned to the land in rain or as "dry deposition." These N inputs are most significant in industrialized areas where they can be as much as 60 kg N/ha/year, but inputs are as low as 5 kg N/ha/year in rural areas that do not have large animal feedlots or dairies.

Nitrogen Fertilizers

Table 3.9 gives the more important water-soluble N fertilizers used in viticulture. Fertilizers such as Nitram (NH_4NO_3) and urea (($NH)_2CO$) supply only N ("single" fertilizers), whereas others such as KNO_3 and ($NH_4)_2SO_4$ supply more than one nutrient and are called multinutrient fertilizers. Mixed or blended fertilizers are made by mixing single or multinutrient fertilizers and are usually identified by their N-to-P-to-K ratio. For example, the mixed

Table 3.9 Forms of Soluble Nitrogen Fertilizer Suitable for Vineyards

Compound	Chemical formula	Nitrogen content (%)	Comments
Potassium nitrate	KNO_3	13	KNO_3 and $Ca(NO_3)_2$ are very
Calcium nitrate	$Ca(NO_3)_2$	15.5	soluble; suitable for fertigation but more expensive per kilogram N than other forms
Ammonium sulfate	$(NH_4)_2SO_4$	21	Supplies N and S in soluble forms; oxidation of the NH_4^+ ions acidifies the soil
Ammonium nitrate (Nitram)	NH_4NO_3	34	NO_3^- is immediately available; NH_4^+ is adsorbed in the soil and oxidized to NO_3^-
Monoammonium phosphate (MAP)	$NH_4H_2PO_4$	11	Water soluble; also supplies P; acidifying
Diammonium phosphate (DAP)	$(NH_4)_2HPO_4$	18	As for monoammonium phosphate
Urea	$(NH_2)_2CO$	46	Very soluble, prone to volatilize if not washed into the soil; acidifying; can be used as a foliar spray
Urea ammonium nitrate solution	$(NH_2)_2CO$, NH_4NO_3 in water	30–32	Has the advantages and disadvantages of urea and NH_4NO_3

fertilizer Horticulture Special has a composition of 10-3-10, meaning 10% N, 3% P, and 10% K.[4]

Water-soluble N fertilizers are supplied as solids or liquids. Fertigation involves applying water-soluble N fertilizers in solution, usually by drip irrigation. There are also liquid organic products, such as seaweed and fish extracts, which have low N concentrations (1%–2% N) but are promoted primarily on the benefits of their organic constituents. Because of the variety of compounds claimed to be present, often in unspecified concentrations, any of these benefits are difficult to confirm scientifically. Although the products are usually applied as foliar sprays, some are also recommended as a soil drench. Because of their low macronutrient concentrations, incomplete absorption through the leaves, and spray drift, it is advocated by vendors of these products that they be applied several times during a growing season, which adds to the cost of their use. Losses may also occur when more concentrated fertilizers are applied directly to the soil as solids or by fertigation, as discussed in the next section.

[4] In the fertilizer trade, analyses are often quoted as the ratio of N to P_2O_5 to K_2O, but for simplicity all analyses are given as elemental percentages in this book.

Potential Losses from Nitrogen Fertilizers

The uptake of fertilizer N applied in a single season is usually less than 50%. Nitrate from $Ca(NO_3)_2$ and NH_4NO_3 is most readily absorbed. Although some of the NH_4-N from NH_4NO_3 and urea is immobilized by microorganisms, most is oxidized to NO_3^-. Much of the NO_3-N that is not taken up by the vines or a mid-row cover crop is lost by leaching or denitrification, the latter occurring when a soil becomes waterlogged (see "What Causes Poor Drainage?" chapter 4).

Not only is N lost as NO_3^- through leaching, but nitrification followed by leaching also acidifies the soil (see box 3.2). Potentially the most acidifying N fertilizers are those in which all N is present as NH_4-N, such as in urea and $(NH_4)_2SO_4$. Less acidifying is NH_4NO_3, whereas $Ca(NO_3)_2$ and KNO_3 have no acidifying effect at all. This is one reason why $Ca(NO_3)_2$ is favored as an N fertilizer in vineyards on acid to neutral soils, especially when applied by drip irrigation. Potassium nitrate is not recommended if the soil already supplies adequate K.

Urea applied to moist soil rapidly hydrolyzes to produce NH_4^+ ions and NH_3 gas that is lost by volatilization. High temperatures and wind accelerate NH_3 loss, and, in extreme cases, up to 50% of the N in surface-applied urea can volatilize. Granules of urea or NH_4NO_3 are best placed a few centimeters into the soil or dissolved in irrigation water to avoid NH_3 loss. However, if too much irrigation is applied, the dissolved urea may leach to greater depths before it hydrolyzes to NH_4^+ ions, in which case the acidity that develops subsequently through nitrification is difficult to correct because it is inaccessible to surface-applied lime.

Controlled-release fertilizer or slow-release fertilizer can be used to reduce N losses. Controlled-release fertilizers, such as sulfur-coated urea (SCU, 31%–38% N), polymer-coated SCU (<30% N), and urea products with a urease inhibitor (46% N), are synthetic compounds for which nutrient release is slow and well controlled. The slow-release fertilizers are synthetic or natural products for which nutrient release is slow but not necessarily well controlled, because it depends on microbial action in the soil. Examples are blood and bone (5%–6% N) and hoof and horn meal (7%–16% N). Both controlled-release fertilizers and slow-release fertilizers are sparingly soluble materials that do not produce the localized high concentrations of NH_4^+ and NO_3^- that predispose to N loss.

When to Apply Nitrogen Fertilizer

The timing and amounts of N influence vine vigor and canopy development, bud fruitfulness, juice N, and N storage in woody tissues. Figure 3.13 shows the periods of N uptake or remobilization within the vine in relation to its stage of growth.

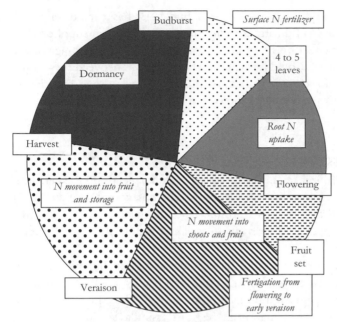

Figure 3.13 Grapevine nitrogen (N) uptake and N fertilizer application in relation to physiological stages of growth and development. (Redrawn from Goldspink and Howes, 2001.)

Root growth lags behind shoot growth from bud burst to the fourth or fifth leaf stage. During this period, N stored in the roots and trunk is mobilized and there is little uptake from the soil. After that, N uptake from the soil predominates as root growth accelerates to reach a peak midway between flowering and veraison. Thus some N should be supplied to the vine at or soon after bud burst by surface application, because the soil is normally too moist for fertigation then. Depending on the N status of the vines, more N fertilizer can be applied, preferably by fertigation, after flowering and up to veraison. An adequate N supply during this period has the most influence on juice N and, ultimately, the success of fermentation (see "Balancing the Nitrogen Supply").

Bunches are the main accumulators from veraison to harvest, with most of the N coming from remobilization within the vine and only a minor portion coming from the soil. Particularly in warmer regions, there is a postharvest period of root growth during which N is taken up from the soil. During this period, much N is also translocated from leaves and shoots to the roots and trunk, where it becomes a reserve for early shoot growth the following season and has an important influence on subsequent vine fruitfulness. If the N status has been well maintained up to harvest, fertilizer is not needed postharvest when there is a greater risk of nitrate leaching by autumn and early winter rains.

In summary, N applied at or soon after bud burst has a major effect on vine growth and crop load during the current season. Nitrogen supplied after fruit set has a major effect on berry N concentration and on N storage in the trunk for the following season. Normally, the fertilizer split between these times would be 1:1, but winegrowers can adjust this according to their experience. Table 3.10 summarizes recommendations to provide adequate N to vines with minimal N losses.

Balancing the Nitrogen Supply

The amount of N fertilizer required during one season depends on the net effect of N inputs (from soil mineralization, legume cover crops, compost, and rain), N removed in harvested fruit, whether prunings are mulched and returned to the soil, and any losses. Appendix 3 gives examples of how the N balance and N fertilizer required are calculated.

Adequate N is important not only for yield but also for maintaining an optimum berry N concentration for fermentation. Yeast strains vary in the efficiency with which they use organic and mineral N for growth. The critical measure of berry N for satisfactory fermentation is the yeast assimilable nitrogen (YAN), comprising NH_4-N and amino acid N (excluding proline). Juice YAN should be in the range of 200 to 480 mg N/L, depending on the yeast strain, with the optimum around 250 mg N/L for red varieties and 250 to 350 mg N/L for whites.

Table 3.10 **Recommendations to Minimize Nitrogen Losses in Vineyards**

Recommendation	Expected outcome
Do not apply soluble N fertilizers after harvest	Prevents NO_3^- being leached by autumn and winter rains
Do not fertigate before flowering	Prevents leaching when soil is still wet from winter
Incorporate urea into the surface soil or wash in with irrigation water (do not apply too much water before the urea has been converted to NH_4^+ ions, which are not readily leached)	Minimizes NH_3 loss by volatilization, especially in warm to hot regions; avoids soil acidification at depth
Use a legume crop as green manure in the mid-rows	Provides a slow release of mineral N to the vines as it decomposes; useful on sandy soil
Use a controlled release fertilizer (CRF) or slow release fertilizer (SRF) in-row	Provides a slow release of mineral N; useful on sandy soil; some slow-release fertilizers are acceptable for organic viticulture
Apply sufficient N to balance losses and removals in product (see "Balancing the Nitrogen Supply" in this chapter)	Provides adequate N for growth but prevents excess mineral N building up in the soil

If YAN is too low, the fermentation rate is sluggish and may become "stuck." Unpleasant odors from hydrogen sulfide (H_2S) can be produced. Although wine-makers can correct this problem by adding monoammonium or diammonium phosphate to the must, it is preferable for the grapes to have an adequate N concentration coming into the winery.

Conversely, if YAN is too high, fermentation is too rapid and poor-quality wine is produced, particularly in the case of red grapes. Residual protein in the must after fermentation causes haze in the wine. Especially when coupled with plentiful water, too high an N supply to the vine leads to excess vigor, a topic considered again in "Managing Soil Water with Irrigation," chapter 4.

Phosphorus Cycling and Phosphorus Fertilizers

As illustrated in figure 3.14, P cycling in a vineyard is similar to N cycling with the following differences:

1. Phosphorus inputs from the atmosphere are negligible.
2. The chemistry of phosphate ions is more complex than N because these ions are adsorbed onto clays and Fe and Al oxides, and P can also form insoluble precipitates. The term "P fixation" encompasses these processes of sorption and precipitation. One consequence of P fixation is that there is normally little water-soluble P in the soil, except around dissolving fertilizer granules, and P does not readily leach from the soil in the way that NO_3^- does.
3. Because of low water-soluble P, vines benefit from mycorrhizal symbioses, as discussed earlier in "Mycorrhizas and Nutrient Uptake."

Soil testing for P aims to measure the amount of available or "labile" P, comprising soluble P and that fraction of solid-phase P that is readily released into solution. Water-soluble P fertilizers augment the labile P, but over time this P reverts to insoluble, less available forms. Thus even though P is not lost by leaching or as a gas, fixation results in an ongoing need for P fertilizers, compost, or manure to maintain healthy vines, unless the soil is naturally rich in P minerals. Because they contain much Fe and Al oxide and hence have a large fixation capacity, soils such as red loams require more P than other soils. Also, more P fertilizer is required at vineyard establishment than for the maintenance of mature vines. The maintenance application can be calculated from fruit yield and the assumption of 0.6 kg P removed per tonne of fruit, with an adjustment for P fixation. Most commercial laboratories can measure a soil's "P sorption" index, which is a surrogate for its P fixation capacity and can be used to adjust the P fertilizer recommendation up or down for either establishment or maintenance.

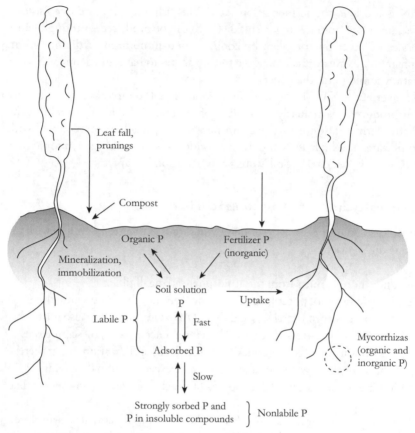

Figure 3.14 The phosphorus (P) cycle in a vineyard. (White, 2003)

Table 3.11 shows the main forms of P fertilizer for use in vineyards. Rock phosphates are of special interest because, being a natural product, they are acceptable in organic viticulture. A commercial rock phosphate should be "soft" rather than "hard," finely ground, and used only on soils of $pH(CaCl_2)$ of 5.5 or less (see box 3.6) to maximize its effectiveness. Because they are slow-release fertilizers, rock phosphates are most effective if mixed into the soil by plowing or ripping at vineyard establishment. Phosphate ions are very immobile in all but sandy soils low in Fe and Al oxides, so even water-soluble fertilizers such as superphosphate are more effective if mixed into the soil by cultivation. Alternatively, soluble P can be supplied by fertigation.

Other Macronutrient Fertilizers for Vineyards

Calcium, Magnesium, and Potassium

Calcium is supplied in fertilizers such as $Ca(NO_3)_2$, single superphosphate, triple superphosphate, and rock phosphate, as well as in various liming materials (see table B3.6.1). Gypsum also supplies Ca and, because it is more soluble than

Table 3.11 **Forms of Phosphorus Fertilizer Suitable for Vineyards**

Compound	Chemical formula	Phosphorus content (%)	Comments
Single superphosphate (SSP)	$Ca(H_2PO_4)_2.H_2O$, $CaSO_4.2H_2O$	9	80%–90% water soluble; remainder mostly soluble in neutral ammonium citrate; supplies P and S
Triple superphosphate (TSP)	$Ca(H_2PO_4)_2.H_2O$	19–21	80% water soluble; 15% citrate soluble; only traces of S
Monoammonium phosphate (MAP)	$NH_4H_2PO_4$	21–26	>90% water soluble, supplies N and P; suitable for fertigation
Diammonium phosphate (DAP)	$(NH_4)_2HPO_4$	20–23	80% water soluble and 20% citrate soluble, supplies N and P; suitable for fertigation
Rock phosphate	$Ca_{10}(PO_4)_6(F,OH)_2$ with variable SiO_2, $CaCO_3$, and Fe, Al oxide impurities	6–18	Water insoluble; citrate solubility depends on whether the rock is classed as "soft" or "hard"

limestone, is useful for counteracting acidity in the subsoil caused by excess exchangeable Al^{3+} (see box 3.6). Gypsum is also used to improve soil structure, especially in sodic soils, as discussed in "Leaching, Salinity, and Sodicity Control," chapter 4.

Magnesium is contained in dolomite, a limestone in which part of the Ca has been replaced by Mg. Magnesium can also be supplied as Epsom salts ($MgSO_4.7H_2O$), which are soluble and can be applied as a foliar spray. Because many Australian soils have high subsoil concentrations of exchangeable Mg^{2+}, an analysis of the 0- to 15-cm depth only may be misleading; petiole analysis is the best indicator of Mg deficiency.

Table 3.12 shows the common Mg and K fertilizers for vineyards. Although potassium sulfate (K_2SO_4) is more expensive than potassium chloride (KCl), it is preferred in salt-affected soils in which Cl toxicity is a potential problem. Potassium sulfate, like KNO_3, also has the advantage of supplying another major nutrient. These three K compounds are soluble and can be applied by fertigation.

Sulfur

Sulfur as elemental sulfur (S) is supplied in sprays for powdery mildew. Sulfur also occurs as the SO_4^{2-} ion in fertilizers such as gypsum (14% S), single superphosphate (11% S), $(NH_4)_2SO_4$ (24% S), and K_2SO_4 (18% S). Additional sulfate-S comes from the atmosphere in rain or as dry deposition, especially in

Table 3.12 **Common Magnesium and Potassium Fertilizers for Vineyards**

Compound	Chemical formula	Element content (%)	Comments
Magnesium[a]			
Epsom salts	$MgSO_4.7H_2O$	10	Water-soluble form of Mg fertilizer; used as a foliar spray at rates up to 10 g/L
Potassium			
Sulfate of potash	K_2SO_4	41	Supplies K and S; preferred on saline soils
Potassium nitrate	KNO_3	38	Very soluble, supplies K and N; not to be used if vine N status is high; used as a foliar spray at rates up to 10 g/L
Muriate of potash	KCl	50	Very soluble in water; granular and easier to handle than K_2SO_4; not suitable as a foliar spray because of Cl content

[a] Magnesium is also supplied in some liming materials (see table B3.6.1).

industrialized regions. Spray sulfur is supplied as a wettable powder with more than 80% of particle sizes less than 125 µm and is easily mixed with water. Rain washes the fine S particles from leaves into the soil where they must be oxidized to SO_4^{2-} before being absorbed by vine roots. The oxidation of S is slow and produces some sulfuric acid (H_2SO_4) that contributes to soil acidification. With these S inputs, specific applications of S fertilizers are rarely required in vineyards.

Micronutrient Fertilizers

Micronutrients are commonly applied as foliar sprays, for the reason that elements such as Fe, Cu, Zn, and Mn are strongly bound by clays and organic matter and also form insoluble precipitates at neutral to alkaline pHs. For example, grapevines are susceptible to Fe deficiency in chalk and limestone soils, a condition called lime-induced chlorosis (see figure 3.7A). Susceptibility to lime-induced chlorosis of vines on own roots or rootstocks is discussed in "Other Rootstock Attributes," chapter 5.

Copper, Zn, and Mn are present in many fungicidal sprays, which, if used, remove the need for dedicated foliar applications. The effectiveness of a spray depends on how easily the nutrient penetrates a leaf and its mobility within the tissue. A surfactant included in the spray improves leaf wetting and nutrient penetration.

Table 3.13 gives examples of inorganic salts or chelates that supply micro-nutrients. Natural chelates comprising Fe, Mn, or Zn complexed with a lignosulfonate are marketed on the basis of better absorption through leaves or from the soil than sulfates. Although this may be true for a soil application, it is not so for a foliar application because the larger chelated molecule is less easily absorbed through the leaf cuticle. Seaweed extracts containing micro- and macronutrients, vitamins, hormones, and amino acids are also marketed for use as foliar sprays or soil drenches. They may provide benefits in addition to their nutrient content because of growth-promoting compounds, but this extra benefit is hard to quantify.

Because seaweed extracts and natural chelates cost more than the inorganic compounds, the latter are preferred provided that "spray-grade" forms are used (with negligible impurities that might cause leaf damage) and correct application procedures are followed. Fifty percent of the nutrient may be lost through spray drift, and up to 50% of the remainder may not be absorbed into the leaves. Although this is important in the short term, remember that residual nutrients on leaves will eventually reach the soil through rain wash and leaf fall, so in the longer term a spray-on micronutrient can build up in the topsoil. Such has been the case in many Bordeaux vineyards where Cu sprays have been applied to combat downy mildew for many years.

Table 3.13 **Examples of Micronutrient Fertilizers**

Fertilizer	Element content (% dry weight)	Comments
Borax	11 B	Soluble in hot water; apply in autumn to avoid phytotoxicity and allow for translocation in the tissues
Copper sulfate	26 Cu	Soluble; also supplies some S; not required if Cu fungicidal sprays are used
Copper chelate	9–13 Cu	Content varies depending on whether it is a natural or a synthetic chelate
Iron sulfate	20 Fe	Soluble, also supplies some S
Iron chelate	6–10 Fe	Content varies depending on whether it is a natural or a synthetic chelate
Manganese sulfate	25 Mn	Soluble, also supplies some S
Zinc sulfate	22 Zn	Potentially phytotoxic if concentration is too high; "neutral Zn" ($ZnSO_4.Zn(OH)_2.CaSO_4$) is safer
Zinc chelate	14 Zn	May be a natural or synthetic chelate
Sodium molybdate	39 Mo	Soluble

Compiled from Goldspink and Howes (2001) and Christensen (2005).

Depending on vine density, recommended spray concentrations are in the range of 1 to 5 g/L in spray volumes of 200 to 1000 L/ha, giving application rates from 0.1 kg/ha for Mo to 0.33 kg/ha for B to 1 kg/ha for Zn. If applied directly to the soil, the rate would be 1 to 7 kg of the element per ha (to be applied once in five to six years). Box 3.8 shows how to calculate the amount of either a macro- or micronutrient fertilizer to be applied per hectare or per meter of vine row.

Box 3.8 How to Calculate Fertilizer Application Rates

The amount of fertilizer required (the "rate") to supply a recommended amount of a nutrient is calculated as follows:

$$\text{Fertilizer rate (kg/ha)} = \frac{\text{kg nutrient per ha}}{\text{nutrient (\%) in the fertilizer}} \times 100 \qquad \text{(B3.8.1)}$$

So if the recommendation is for 30 kg N/ha and the fertilizer chosen is NH_4NO_3 (34% N), the rate required is

$$\frac{30}{34} \times 100 = 88 \text{ kg/ha} \qquad \text{(B3.8.2)}$$

Rates for micronutrient fertilizers applied to the soil are calculated the same way, except that these rates are much lower than for N, P, and K.

Usually fertilizer is applied to the vine rows, in which case the amount required per row will depend on the row spacing. Suppose the recommended rate is y kg/ha and the row spacing is x meters. The total length of rows per hectare will be $10,000/x$ (meters) and the amount of fertilizer required per meter of row is given by

$$\left(\frac{yx}{10000} \right) \text{kg} \qquad \text{(B3.8.3)}$$

The equation to convert from a fertilizer rate in kilograms per hectare to an amount (grams) per square meter is

$$\text{g fertilizer per m}^2 = \frac{\text{kg fertilizer per ha}}{10} \qquad \text{(B3.8.4)}$$

Multinutrient and Mixed Nutrient Fertilizers

Winegrowers are offered a bewildering range of specialist multinutrient and mixed fertilizers in solid and liquid form. Table 3.14 gives a few Australian examples. Some of these fertilizers supply the macronutrients N, P, and K in various ratios, with or without a suite of micronutrients (commonly labeled "trace elements"), while others supply a mixture of micronutrients only (either as inorganic salts or complexed with organic agents such as fulvic acid or a ligno-sulphonate). Yet others primarily supply organic compounds such as molasses, humic and fulvic acids, and unquantified concentrations of vitamins and plant growth hormones derived from materials such as kelp (seaweed) and fish emulsion. Beyond their content of the essential nutrients, the value of these extra compounds, at the relatively low concentrations recommended, is scientifically dubious. Furthermore, the commercial micronutrient products are "shotgun" mixtures, which are recommended irrespective of any identification of a deficiency by plant analysis. Although some winegrowers are convinced that the use of such multinutrient and mixed fertilizers is beneficial for their vines, others will

Table 3.14 Examples of Propriety-Brand Multinutrient and Mixed Nutrient Fertilizers

Fertilizer product[a]	Analysis (%)	Comments (from product descriptions)
Baseline Plus	12-5-14 (N-P-K)	Source of N, P, K, and unspecified chelated trace elements and biological stimulants
TE 6 Plus	2.6 N, 0.1 K, 4.2 S, 2.4 Mg, 3.1 Mn, 3.1 Zn, 0.47 Cu, 0.02 Mo, 0.21 B, 0.73 Fe, 0.05 Co, 0.5 fulvic acid	A focus on Mg, Mn, Zn and Cu with fulvic acid as a chelating agent "to benefit plant permeability"
QuadSHOT	0.4 N, 2.4 P, 2.9 K, 0.2 Ca, 0.002 B, 0.001 Zn, 7.3 fish emulsion, 7.3 kelp, 7.3 molasses, 6.0 humic acid, 0.24 fulvic acid	Organic additives and stimulants aid nutrient cycling, root growth, soil biota; the mixture increases soil *CEC* and "improves saline and sodic soils"
Smartrace products	N, P, K, S, Ca, Mg, Fe, Mn, Cu, Zn, B, and Mo in various combinations and different concentrations	Free-flowing liquid solutions containing complexing lignosulphonates; "trace elements are released at a controlled rate"
Gripper products	Micronutrients, Ca, and Mg in various combinations and different concentrations	Ethylenediaminetetraacetic acid complexes the nutrient elements, which "will easily penetrate the leaf cuticle"

[a] The product range is likely to change over time.

find that the extra expense of these fertilizers compared with conventional fertilizers is not justified. The next section discusses some of these organic products in more detail.

Organic Viticulture

What Does It Mean?

Organic viticulture aims to produce quality wine without the use of synthetic fertilizers or chemicals. To the maximum extent possible, an organic system must operate as a closed system, with external inputs used only on an "as-needed" basis. The broad term organic includes biodynamic (BD) and biological systems, but BD viticulture has requirements additional to a simple organic system. These requirements include timing vineyard operations to coincide with phases of the moon and using special preparations (numbers 500 to 507) that are not fertilizers and are applied either to the soil (e.g., preparation 500, to stimulate soil biological activity) or sprayed on the canopy (e.g., preparation 501, to attract the "light forces" and aid photosynthesis and the vine's resistance to disease). In both cases, these preparations must be applied at regular intervals during the growth season because of the extremely low concentrations of the active ingredients. This is especially true of preparation 500 when applied to soil because of the pressure on the "inoculum" to survive and multiply in competition with an overwhelming number of native soil microorganisms.

BD systems of farming have evolved from a series of lectures given by the Austrian polymath Rudolph Steiner in 1924. Just as there is a range of viticultural practices in conventional and organic winegrowing, there are different levels of adoption of BD practices, depending on the strength of a winegrower's belief in the philosophical concepts and "cosmic principles" espoused by Steiner (Lachman, 2007). These levels have been summarized by Max Allen in Australia, a wine writer who is an enthusiastic proponent of BD winegrowing (see www.redwhiteandgreen.com.au).

Traditionally strong in France, organic viticulture (including BDs) is expanding worldwide and now accounts for approximately 5% of western European production (Johnston, 2013). While certified organic production in Australia is only about 0.5%, as indicated in chapter 1, Organic Winegrowers of New Zealand have the aim of 20% of that country's production being organic by 2020. A vineyard and its wine must be certified by an accrediting agency to be labeled organic. The overarching global agency is the International Federation of Organic Agriculture Movements (www.ifoam.org), including certifying agencies such as Ecocert International (www.ecocert.com), recognized in the European Union, and the National Organic Program of the U.S. Department of Agriculture (www.ams.usda.

gov). In Australia, the peak body is the Organic Federation of Australia (www.ofa. com.au), whose members apply the National Standard for Organic and Biodynamic Produce developed by Standards Australia under the auspices of the Department of Agriculture (www.daff.gov.au/agriculture-food/food/organic-biodynamic).

To be certified as organic, a vineyard must be established on organic principles from the beginning or must undergo a three-year conversion period, although after two years the wine can be labeled as "in conversion." Certified organic production means that grapes are grown without insecticides, herbicides, fungicides, and chemical fertilizers, other than those approved by the certifying agency, using cultural practices that minimize adverse effects on the soil and wider environment. Genetically modified plant material cannot be grown, nor can genetically modified yeast be used in fermentation. Chapter 6 includes an appraisal of the overall advantages and disadvantages of organic viticulture and a summary of any observed effects on wine quality.

Complementing the organic movement, winegrowers in the major wine-producing countries are encouraged to adopt an integrated production system, discussed in chapter 6. Briefly, such a system involves minimizing pesticide and fungicide sprays through integrated pest management to encourage a balance between natural predators and pests and practicing good canopy management to avoid disease problems. Fertilizer recommendations are still based on soil testing, and organic manures, compost, and mulch are used wherever possible. Mid-row cover crops or permanent grass swards are advocated to improve soil structure and provide a favorable habitat for beneficial organisms and predators of pest insects.

Cultural Practices for Organic Viticulture

In organic viticulture, weeds are controlled by mechanical cultivation instead of herbicides. Mowings from a mid-row cover crop can be thrown under the vines to serve as mulch. A cover crop containing legumes can be cultivated into the soil as "green manure." The slow release of mineral N as the green manure decomposes minimizes the chance of any NO_3^- formed being leached before the vine can absorb it.

Farm animal manures must be composted or followed by at least two green manure crops after application. Biosolids derived from composting sewage sludge are generally prohibited unless specifically exempted by the certifying agency. The most commonly used manures in vineyards are poultry manure, followed by sheep manure. Composted manures should be pelleted to produce an odorless product in which N is at first stabilized by microbial immobilization but the C-to-N ratio is reduced to less than 25 so that net N mineralization occurs in the soil (see figure 3.5). Composted manures and biosolids (if permitted) provide macronutrients such as N, P, K, and S; micronutrients, trace elements, and salts;

thus winegrowers must take care that the amounts applied are not excessive for vines. This is particularly so for pig manure, which has relatively high concentrations of Cu, Zn, and Mn.

Composted green garden waste from cities is also used in vineyards, as is compost made from pomace (pressed skins and seeds from a winery, called marc in Australia; figure 3.15). After steam distillation to remove residual alcohol and most of the tartaric acid, a product called spent marc is available to farmers for use as a soil conditioner, stock feed, or biofuel. However, marc or spent marc contains significant amounts of K (range 1% to 2% of the dry weight) that can cause a problem for wine quality when used as a soil conditioner in vineyards. For a general improvement in soil fertility and structure, composted manures and compost can be spread through the mid-rows at rates up to 10 t/ha and worked into the soil in winter when the vines are dormant. For a more targeted application, compost is spread under the vines to a depth of at least 5 cm (a rate of around 2 t/ha), serving as a mulch that suppresses weeds, reduces soil temperatures, and slows evaporation. Compost tea, an organic-rich liquid that drains from a compost heap, is sometimes applied as a foliar spray, but its benefits are not proven. Composting is discussed further in "Manures and Composts," chapter 5.

Table 3.15 lists examples of manures and organic fertilizers acceptable in organic viticulture. Some are by-products from animal processing, such as hoof and horn meal, and others are plant-derived products such as seaweed extracts that are recommended as foliar sprays. Because of their diverse sources, the commercial products have a wide range of nutrient contents. As with BD preparations, nonspecific benefits are claimed for some of these products, such as the fish meal and seaweed extracts, in strengthening a plant's disease resistance and, when used as soil drenches, stimulating soil biological activity. Certainly, research has shown

Figure 3.15 An example of well-composted grape marc (pomace) in an organic vineyard in the Upper Goulburn region, Victoria, Australia.

Table 3.15 Examples of Manures and Organic Fertilizers for Use in Organic Viticulture

Material	N	P	K	Cu	Zn	Mn	Comments
Pelleted poultry manure[a]	3–4.6	1–2.8	1.2–1.7	30–170	220–300	300–580	Composted; variable depending on age and content of straw or sawdust
Blood and bone[a]	7–8	4–5					Slow release of N and P
Hoof and horn meal[a]	7–16						Slow release of N
Fish meal[b]	12	1.4	7				Applied as a liquid containing 50% solids
Compost[a]	1.6–1.9	0.8–0.9	0.8–1.2	100–200	700–900	300–400	Variable depending on age and material used
Seaweed extract (e.g. Tasmanian Bull Kelp; seaweed from Nova Scotia)[b]	0.75–2	0.15–1.3	1.2–16	1–10	25–75	8–12	Usually applied as a foliar spray but can be used as a soil drench; boron content is high at 75–150 mg/kg; contains plant growth hormones and amino acids

[a] Units for macronutrients (percent) and micronutrients (milligrams per kilogram) on a dry matter basis.

[b] Units for macronutrients (percent) and micronutrients (milligrams per kilogram) by weight of product containing varying amounts of liquid.

Compiled from Goldspink and Howes (2001), White (2003), and product pamphlets.

that many soil indicators, such as organic C content, microbial biomass, and species diversity, are better under organic or BD systems than conventional systems. However, fruit yields are less in the former, primarily because nutrient inputs are less and in some cases because of weed competition. Importantly, the effects on soil biology in organic and BD systems are only comparable if inputs of organic materials are the same in both. Moreover, comparable benefits for soil biology can be achieved under conventional viticulture if inputs of organic materials are adequate. As discussed in "Organic, Biodynamic, and Conventional Viticulture," chapter 6, the effects of organic or BD systems on wine quality compared to conventional systems have been found to be small and inconsistent.

Table 3.16 lists the range of minerals from natural sources that are acceptable as fertilizers or soil amendments in organic viticulture.

Precision Viticulture for Better Soil Management

Precision viticulture embodies the concept of measuring soil variation with a dense spatial array of observations and using this information to plan the layout, management, and harvesting of a vineyard most effectively. Whereas figure 3.10 illustrated the point that, although the P status of a block may be deemed adequate based on one composite sample (figure 3.10A), the spatial pattern of variation shown in figure 3.10B reveals that more than half the vines are deficient or marginal in P (figure 3.10C). Clearly, correcting the P deficiency through targeted fertilizer application would improve the performance of this vineyard block. However, the cost of sampling and analysis (190 samples in this case) by standard "wet" chemical methods would be prohibitive for a commercial vineyard.

Soil properties sensed by rapid, in-field methods have the potential to be surrogates for expensive laboratory measurements. The EM38 measurements described in chapter 2 are an example of a surrogate method for assessing the variation in soil properties such as salinity, depth, or clay content. Midinfrared diffuse reflectance spectroscopy is another inexpensive and rapid surrogate method for measuring soil properties at a high spatial resolution. After calibration against a database of a soil property values, measured by a standard method, midinfrared diffuse reflectance analysis can be applied to new sites. Figure 3.16 is an example of this method used to map the soil $CaCO_3$ content of a vineyard block. Alternatively, the method can be calibrated using a "training" data set derived from a standard analysis of a limited number of samples, covering the full range of variability at a particular site.

However, the present interest in precision viticulture techniques in commercial viticulture is focused on mapping yield variation in relation to soil variability, which can be caused by several factors, some of which are amenable to management and

Table 3.16 Minerals from Natural Sources Acceptable as Fertilizers or Soil Amendments in Organic Viticulture

Material	Function	Comments
Rock phosphate, phosphatic guano	Supplies P, Ca	Slow release; must be applied to soil of pH<6 (water) or 5.5 (CaCl$_2$) and be finely ground to be effective; small liming effect
Elemental S	Supplies S	Wettable powder used for control of powdery mildew
Wood ash	Supplies K, Ca, and Mg	Timber must not have been chemically treated
Rock dust	Supplies Ca, Mg, some K, and micronutrients	Very slow release; must be finely ground
Rock potash and sulfate potash (K$_2$SO$_4$)	Supplies K, S	Mined product
Epsom salts (MgSO$_4$)	Supplies Mg, S	Natural product
Lime (CaCO$_3$)	Raises soil pH, supplies Ca	See box 3.6 for specifications
Gypsum (CaSO$_4$.2H$_2$O)	Improves soil structure, supplies Ca	Must be from a natural source
Dolomite (Ca,Mg) CO$_3$	Raises soil pH, supplies Ca, Mg	See box 3.6 for specifications

Compiled from Goldspink and Howes (2001) and the National Standard for Organic and Biodynamic Produce (2009).

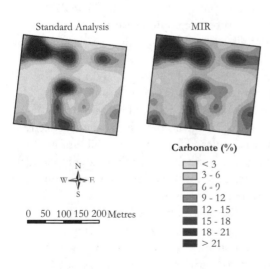

Standard Analysis   MIR

N
W——E
S

0   50 100 150 200 Metres

Carbonate (%)

- < 3
- 3 - 6
- 6 - 9
- 9 - 12
- 12 - 15
- 15 - 18
- 18 - 21
- > 21

Figure 3.16 Good agreement shown between maps of soil calcium carbonate (CaCO$_3$) content (5 to 15-cm depth) in a small Coonawarra vineyard, either measured by a standard method or predicted from midinfrared diffuse reflectance. The latter was calibrated using a data set derived from the analysis of a large number of Australian soils. (Courtesy of Dr. Robert Bramley, CSIRO Ecosystem Sciences, Adelaide, South Australia.)

some not. Data are obtained with a yield monitor attached to a mechanical harvester and imported into a geographic information system. If the pattern of yield variation is consistent from year to year, blocks of vines can be identified for which soil management can be tailored to achieve desired outcomes, according to the style of wine produced and the price point. As more data are recorded from year to year, under different seasonal conditions, the information for decision making can be revised so that uncertainty in the predicted outcomes is decreased. Proffitt et al. (2006) give more details of precision viticulture and its applications in vineyards, a topic developed further in "Managing Natural Soil Variability in a Vineyard," chapter 6.

Summary Points

1. Grapevines rely on the soil to supply the following essential elements: N, P, S, Ca, Mg, K, Cl, Fe, Mn, Zn, Cu, B, and Mo, which are taken up as inorganic ions (nutrient ions) from the soil solution. The elements from N through to Cl are called macronutrients because vines require them in large concentrations (>1000 ppm) compared with the elements from Fe to Mo—the micronutrients, which are required in low concentrations (<1000 ppm). Carbon and O are supplied as CO_2 through photosynthesis and H and O as water.

2. Most of the nutrient ions are held by adsorption onto charged clay, oxide, and organic matter surfaces, which affects their availability to vines. Nutrients such as P, Ca, Fe, Mn, Zn, and Cu can also form insoluble precipitates. Because of the strong adsorption of P by clays and oxides, and its tendency to form insoluble precipitates, "available" P in soil decreases markedly with time, a process commonly called P fixation. Arbuscular mycorrhizas, a symbiotic association between a plant root and soil fungus, can improve the P supply to vines in P deficient soil.

3. Although NH_4^+ and NO_3^- are the forms of N taken up by vines (the available N), 99% or more of soil N exists in an organic form. Similarly, a significant proportion of soil P and S is in organic forms, and the availability of N, P, and S is influenced by the balance between mineralization (conversion to a mineral form) and immobilization (incorporation into microorganisms and thence into humic compounds). Iron and the other micronutrients except B and Mo also form complexes of limited availability with soil organic matter.

4. Nutrient deficiencies and toxicities can be identified from visual symptoms, plant analysis, and soil testing. The grapevine, being a perennial, is particularly suited to plant analysis for both macro- and micronutrients. Protocols exist to determine which part of a vine to sample, when to sample it, and how many samples to take. Regular plant analysis (every one to two years) is recommended to identify trends in

nutrient status rather than relying on an analysis at one time. Spot samples can be taken at any time up to veraison if a problem suddenly arises.

5. Soil testing is most appropriate at the time of vineyard establishment. Soil tests are available for chemical, physical, and biological properties. The most relevant chemical tests for established vineyards are for soil pH, organic C, salinity (the concentration of soluble salts), and sodicity (related to the exchangeable Na percentage). Appropriate physical tests are discussed in chapter 4 and biological tests in chapter 5.

6. Nutrient deficiencies are corrected through the use of fertilizers (as solids or dissolved in irrigation water or as foliar sprays), green manure (in the case of N), animal manures, and composts. Fertilizers range from single element fertilizers such as urea ($(NH)_2CO$, supplying N), to multinutrient fertilizers such as monoammonium phosphate (NH_4HPO_4, supplying N and P), to blended fertilizers such as mixtures of potassium chloride (KCl) and superphosphate ($Ca(H_2PO_4).CaSO_4$), supplying K, P, Ca, and S. Micronutrients are most effectively supplied as foliar sprays.

7. Many commercial products offered to growers are blended fertilizers with additional organic constituents such as humic and fulvic acids, molasses, and extracts from seaweed or fishmeal. They are promoted not only for their macronutrient content but also for the supply of growth-promoting hormones and vitamins, the efficacy of which is difficult to substantiate. Some of these products contain "shotgun" additions of micronutrients (labeled as trace elements), and vendors recommend their use, irrespective of whether a micronutrient deficiency has been identified.

8. A toxicity may be caused by a nutrient imbalance that can be corrected by adjustments to the fertilizers or manures/composts applied, emphasizing the point that the nutrient content of all added materials must be known. Other forms of toxicity may arise if the soil pH is outside the optimum range of 5.5 to 7.5 (measured in $CaCl_2$ solution). Aluminum toxicity is a potential problem at low pH, whereas B and salt toxicity (resulting from too much NaCl) can occur at high pH. High soil pH, especially in limestone-derived soils, can induce deficiencies of Fe, Mn, Zn, and Cu, but Mo becomes more available at high pH.

9. Organic viticulture, including biological and BD viticulture, aims to be a closed system with external inputs used only as needed. Thus synthetic herbicides, insecticide, fungicides, and chemical fertilizers are avoided, except where approved by a certifying body. Composted manures, preferably poultry or sheep manure, and composted vegetable matter are used in organic vineyards where possible. Provided organic materials are added in sufficient quantities (>10 t/ha) and regularly, they can increase soil organic matter and biological activity in conventional, organic, or BD vineyards.

10. In BD viticulture, special preparations such as 500 and 501 are sprayed, at very low concentrations, on the foliage or applied to the soil ostensibly to strengthen the vine's disease resistance or stimulate the soil biota. When applied to soil, the very low numbers of microorganisms in these preparations are unlikely to survive in competition with the overwhelming numbers of native soil microorganisms; hence the treatment is unlikely to be effective.

11. Precision viticulture provides the tools to determine, at a spatial resolution of 1 to 2 m, the distribution of any soil nutrient deficiencies, and other limiting factors such as salinity. These spatial patterns of variation in soil properties can be matched against geographic information system-based maps of yield or canopy density, acquired over several seasons, thereby allowing more effective vineyard management.

Figure 1.4 A Terra Rossa soil formed on calcrete-capped porous limestone in the Coonawarra region, Australia. Vines are in the background.

Figure 1.8 A deep podzol soil formed on sandstone near Sydney, Australia. Iron complexed with organic matter has been translocated from the bleached horizon, to be deposited as orange-red iron oxide in the subsoil. The pale soil on top is spoil from the pit.

Figure 1.10 The gradational profile of a Brown Earth under vines in the Goulburn Valley region, Victoria, Australia. There is a band of precipitated calcium carbonate at depth.

Figure 1.11 A deep red loam on colluvial basalt in a vineyard in the Willamette Valley region, Oregon.

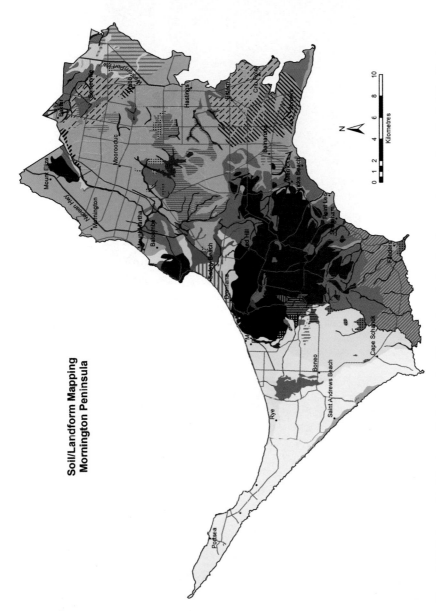

Figure 1.14 Soil map of the Mornington Peninsula region, Victoria, Australia, based on the general-purpose Australian Soil Classification. The map scale is 1:100,000. (Map from the Victorian Resources Online website, hosted by the Victorian Department of Environment and Primary Industries.)

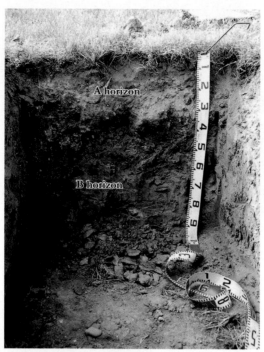

Figure B1.1.1 A vineyard soil in the Gippsland region Victoria, Australia, showing distinct A and B horizons. The A horizon is a bleached sandy loam and the B horizon is a pale orange-mottled clay. There is a thin layer of organic enrichment at the top of the A horizon (A1). The major divisions in the scale are at 10 cm intervals.

Figure B1.3.2 This orange-mottled soil occurred in the subsoil of a duplex soil in the Sunbury region, Victoria, Australia. Note smearing of the moist clay by the excavator.

Figure 2.5 (A) Deep red loam soil in the Heathcote region, Victoria, Australia. The scale is 15 cm. (B) A vineyard soil in the Hawkes Bay region, New Zealand, with an impermeable subsoil that experiences intermittent waterlogging.

Figure 3.7 (A) Iron deficiency in a vineyard in the St. Emilion appellation, Bordeaux region, France. Note yellowing of the younger leaves and marked interveinal chlorosis. (B) Zinc deficiency in grapevines on a high pH soil in the Riverina region, New South Wales, Australia. Leaves are not as severely chlorotic as with Fe deficiency in figure 3.7A.

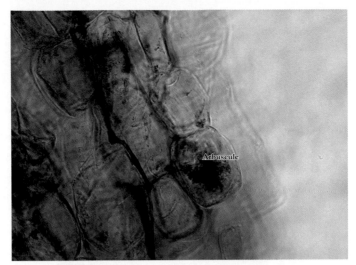

Figure 3.4 Example of an arbuscular mycorrhizal fungus infecting a grapevine root. The branched fungal structure inside one of the host cells is an arbuscule through which nutrients and carbon compounds are exchanged between the host and fungus. (Photo prepared by Dr. Melanie Weckert, National Wine & Grape Industry Centre, Wagga Wagga, Australia.)

Figure 4.15 Cabernet Sauvignon vines showing signs of severe water stress (leaf death) in late summer in the Napa region, California.

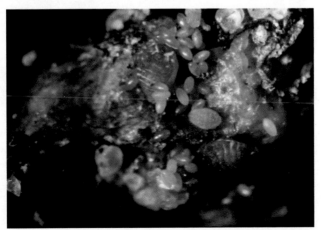

Figure 5.12 Phylloxera adults and eggs infesting a *Vitis vinifera* root. Note the swelling and yellow galls. (White, 2003. Copyright © The State of Victoria, Department of Primary Industries. Reproduced by permission of the Department of Primary Industries and therefore not an official copy. Photo by Greg Buchanan.)

Figure 6.2 A deep red clay loam formed on basalt in a vineyard in the Mornington Peninsula region, Victoria, Australia. The vine roots (painted white) go down to 1 m; the pit depth is 1.2 m.

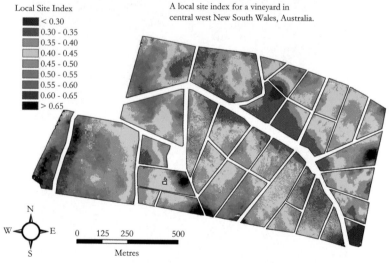

Local Site Index

- < 0.30
- 0.30 - 0.35
- 0.35 - 0.40
- 0.40 - 0.45
- 0.45 - 0.50
- 0.50 - 0.55
- 0.55 - 0.60
- 0.60 - 0.65
- > 0.65

A local site index for a vineyard in central west New South Wales, Australia.

N
W — E
S

0   125   250   500

Metres

Figure 6.8 Map showing the digital *terroirs* for a vineyard in the Cowra region, New South Wales, Australia. The *terroirs* are defined by a local site index incorporating site-specific net radiation, temperature differences due to aspect and height, readily available water, clay-to-silt ratio, and root zone depth. Sites with low local site index are better suited to late varieties because of lower temperatures and frost risk. These sites also have high readily available water and are not suited to vigorous varieties like Shiraz. (Courtesy of Dr. James Taylor, Australian Centre for Precision Agriculture, Sydney, Australia.)

4

Where the Vine Roots Live

Soil Structure

Chapter 3 gives examples of how grapevines, being woody perennials, have the potential to develop extensive, deep root systems when soil conditions are favorable. One of the most important factors governing root growth is a soil's structure, the essential attributes of which are

- Spaces (collectively called the pore space or porosity) through which roots grow, gases diffuse, and water flows
- Storage of water and natural drainage following rain or irrigation
- Stable aggregation
- Strength that not only enables moist soil to bear the weight of machinery and resist compaction but also influences the ease with which roots can push through the soil

The key attributes of porosity, aeration and drainage, water storage, aggregation, and soil strength are discussed in turn.

Porosity

Various forces exerted by growing roots, burrowing animals and insects, the movement of water and its change of state (e.g., from liquid to ice) together organize the primary soil particles—clay, silt, and sand—into larger units called aggregates.

Between and within these aggregates exists a network of spaces called pores. Total soil porosity is defined by the ratio

$$\text{Porosity} = \frac{\text{Volume of pores}}{\text{Volume of soil}}$$

A soil's A horizon, containing organic matter, typically has a porosity between 0.5 and 0.6 cubic meter per cubic meter (m^3/m^3)—also expressed as 50% to 60%. In subsoils, where there is little organic matter and usually more clay, the porosity is typically 40% to 50%. Box 4.1 describes a simple way of estimating a soil's porosity.

Box 4.1 Calculating Soil Porosity

A simple equation for calculating porosity is

$$\text{Porosity} = 1 - \frac{\rho_b}{\rho_p} \qquad\qquad (B4.1.1)$$

In this equation, ρ_p (rho p) is the average density of the soil particles, assumed to be 2.65 Mg (megagram) per m^3. The term ρ_b (rho b) is the soil's bulk density, which ranges from less than 1 Mg/m^3 for soils rich in organic matter to 1.0 to 1.4 Mg/m^3 for well-aggregated loamy soils, and 1.2 to 1.8 Mg/m^3 for sands and compacted subsoils. Thus for the A horizon of a loamy soil with a ρ_b value of 1.33 Mg/m^3, we have

$$\text{Porosity} = 1 - \frac{1.33}{2.65} = 0.5 \qquad\qquad (B4.1.2)$$

that is, the porosity is 0.5 m^3/m^3 or 50%.

To measure a soil's bulk density, take five to six intact cores with steel cylinders, preferably at least 5 cm in diameter and 6 to 10 cm deep. Trim the soil so that the dimensions of the soil core are the same as the cylinder and the soil volume is easily calculated. Dry the cores in an oven at 105°C and weigh each core to obtain the weight of oven-dry (o.d.) soil. The bulk density ρ_b of each core is calculated from the equation

$$\rho_b = \frac{\text{Weight of o.d. soil in the core}}{\text{Volume of soil core}} \qquad\qquad (B4.1.3)$$

Calculate the average value of ρ_b for the several samples taken and substitute this into equation B4.1.2. Note that the soil bulk density in the vine rows and mid-rows is likely to be different because of compaction by wheeled traffic in the latter.

Figure 4.1 A vineyard soil with swelling clays showing cracks on drying.

Total porosity is important because it determines how much of the soil volume water, air, and roots can occupy. Equally important are the shape and size of the pores. The pores created by burrowing earthworms, plant roots, and fungal hyphae are roughly cylindrical, whereas those created by alternate wetting and drying appear as cracks (figure 4.1). Overall, however, we express pore size in terms of diameter (equivalent to a width for cracks). Table 4.1 gives a classification of pore size based on pore function.

Aeration and Drainage

A soil's pore space is normally occupied by water (the soil solution) and air. When all the pores are filled with water, the soil is saturated. As water drains out or evaporates from the surface, gases of the air, predominantly oxygen (O_2) and nitrogen (N_2), enter the pore space and the soil becomes unsaturated. The gases enter the largest pores first, primarily by mass flow, accompanied by diffusion into smaller pores. Normal respiration of soil organisms and plant roots consumes O_2 and produces carbon dioxide (CO_2), and the exchange of these gases between the atmosphere above and a soil's atmosphere is called aeration.

Provided there are some continuous pores in the soil, gas exchange through these pores produces respective concentrations of N_2 and O_2 of 78% and 20% (by volume) in the soil air, similar to the atmosphere. The concentration of CO_2 in soil air normally ranges from 0.1% to 1%, compared to 0.04% in the

Table 4.1 **Relationship among Pore Size, Formative Forces, and Pore Function**

Pore diameter (µm)	Biotic or physical agent	Pore function
5000–500	Cracks resulting from drying; small animals, earthworms, primary plant roots	Aeration and rapid drainage
500–30	Grass roots, small insects	Normal drainage and aeration
30–0.2	Fine lateral roots, fungal hyphae, and root hairs	Storage of "available" water (see "Aeration and Drainage" in this chapter)
<0.2	Wetting and drying of water associated with clay mineral surfaces	Retention of residual or "nonavailable" water

Compiled from Cass et al. (1993) and White (2006).

atmosphere. However, in most soils there are also "dead-end" pores (up to 5% of the total) that remain gas-filled even when the soil appears to be saturated. The O_2 in these pores is soon depleted, with the result that respiration switches from aerobic (in the presence of O_2) to anaerobic (in the absence of O_2) in this pore space, and CO_2 begins to accumulate. Poorly drained soils and those affected by a high water table can be predominantly anaerobic so that gases such as nitrous oxide (N_2O) and methane (CH_4) may be produced. Consistent with total porosity, the air-filled porosity is expressed as a fraction of the soil volume (m^3/m^3) according to the ratio

$$Air\text{-filled porosity} = \frac{\text{Volume of air-filled pores}}{\text{Volume of soil}}$$

Water Storage

About two days after a soil has been thoroughly wetted by rain or irrigation, drainage becomes slow and the soil is said to attain its field capacity (*FC*). To measure the water content at *FC* accurately, the soil surface, which should be bare of plants, must be covered after wetting to prevent evaporation. The water content at *FC* sets the upper limit for stored water available to the vines and any mid-row cover crop. Gradually, as plants extract this water, narrower and narrower pores become filled with air until the water remaining is too difficult to extract and plants wilt. If this condition persists, the permanent wilting point (*PWP*) is reached (see "Redistribution of Water in the Soil" in this chapter). The amount of water held between *FC* and *PWP*—the soil's "available water"—is an important soil property. Box 4.2 summarizes different ways of measuring a soil's water content.

Box 4.2 Ways of Measuring Soil Water Content

Soil water content is most easily measured in the laboratory by weighing a sample of moist soil, drying it at 105°C for 48 hours in an oven and reweighing the sample to measure the weight of oven-dry (o.d.) soil. The loss in weight represents the weight of water, expressed as a percentage of the o.d. soil, according to the equation

$$\text{Gravimetric water content} = \frac{\text{Loss in weight of soil sample}}{\text{Weight of o.d. soil}} \times 100 \quad \text{(B4.2.1)}$$

Another measure of soil water content is the volumetric water content, θ (theta), expressed as m³ water per m³ soil, according to the equation

$$\theta = \frac{\text{Volume of water in the pores}}{\text{Volume of soil}} \quad \text{(B4.2.2)}$$

Volumetric water content can be measured directly in the field (see "Monitoring Soil Water," this chapter). Also, because 1 Mg water occupies 1 m³ at normal temperatures, a gravimetric water content can be converted to a θ value by multiplying by the soil's bulk density.

Values of θ are useful in the vineyard because they give a direct measure of the "equivalent depth" of water per unit area of the soil. For example, consider a soil volume of 1 m³ that has a θ value of 0.25 m³/m³, which is equivalent to a water depth of 0.25 m per m² surface area. As shown in figure B4.2.1, we may visualize this as a depth of 250 mm of water in 1-m depth of soil. Expressed this way, soil water content is directly comparable with volumes of rainfall, irrigation water, and evaporation, all of which are measured in millimeters (per m² of soil surface). A good rule of thumb is that 1 mm of rain or irrigation water is equivalent to 1 liter (L) per m².

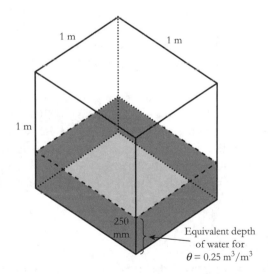

Figure B4.2.1 Equivalent depth of water in 1 m³ of soil of water content θ equal to 0.25 m³/m³. (White, 2003)

(continued)

Box 4.2 *(continued)*

A θ value of 0.25 m³/m³ is the same as 2.5 mm per cm depth of soil. Thus the equivalent depth d (in millimeters) of water in any soil depth z (in centimeters) is given by

$$d = 10 \times \theta \times z \qquad\qquad\qquad (B4.2.3)$$

Table B4.2.1 gives some examples of the use of this equation.

Table B4.2.1 **Some Examples of Equivalent Depths of Water in Soil**

Soil volumetric water content θ (m³/m³)	Soil depth z (cm)	Equivalent depth of water d in soil depth z (mm)
0.25	20	50
0.25	40	100
0.40	20	80
0.40	40	160

At any given time, the sum of the water-filled and air-filled porosities equals the total porosity. When the soil is at its *FC*, the value of air-filled porosity is called the air capacity, or sometimes the drainable porosity. The volume of available water in a 1-m-deep profile (see box 4.2) is called the available water capacity (*AWC*). These two variables—air capacity and *AWC*—can together be used to classify a soil's structural quality, as shown in figure 4.2.

Soil droughtiness increases as the amount of available water decreases, whereas its susceptibility to waterlogging increases as the air capacity decreases. An air capacity of 15% together with an *AWC* of 20% or greater is regarded as being very good. The simple classification of figure 4.2 is useful for vineyard soils because it sets practical limits for satisfactory aeration and available water. Additionally, by taking into account the effect of soil strength on root penetration, it is possible to refine this simple classification of soil structural quality, as discussed in "Soil Strength" later in this chapter.

Soil Aggregation

The size, shape, and consistence of aggregates vary considerably between soil types and often between the topsoil and subsoil of one soil type. For example, figure 4.3 shows a surface soil that is well aggregated, with aggregates mainly

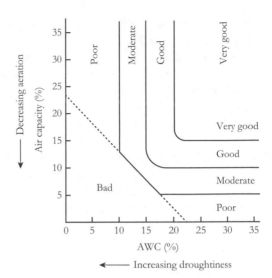

Figure 4.2 A simple classification of structural quality for vineyard soils. (White, 2003)

Figure 4.3 Desirable subangular blocky structure in the topsoil of a calcareous soil in the Tuscany region, Italy. The scale is 10 cm.

blocky but well rounded and between 5 and 20 mm in size. This kind of aggregation is typical of loamy soils under grass. The fine roots of the grass plants bind soil particles together, whereas gums and mucilages produced by the roots and associated bacteria and fungi act as a biological adhesive. The more organic matter is in the aggregates, the darker they appear (see figure 3.6, chapter 3). This kind of aggregation predisposes to not only good drainage and aeration but also adequate water storage.

Although grass roots and organic compounds are the key to good surface soil structure, their effect is less significant in the subsoil because they are less abundant there. Good subsoil structure consists of aggregates that are orientated vertically and are longer than they are broad, with well-defined cracks between and within them. Ideally, the texture range is loam to clay loam to light clay, and the color is golden brown to red, without mottles, showing that the subsoil is well drained and not waterlogged when wet. The dominant exchangeable cation is Ca^{2+}, with iron (Fe) and aluminum (Al) oxides frequently acting as bridging compounds between clay particles. Figure 4.4 shows an example of a well-structured subsoil in a vineyard in the Coonawarra region, South Australia.

Forces at Clay Surfaces

The amount and type of clay minerals have an important effect on aggregate formation and stability. As described in "Retention of Nutrients by Clay Minerals and Oxides," chapter 3, layers within the clay crystals have an overall negative charge that is just balanced by the positive charge of exchangeable cations attracted to the surfaces. The cations available in the soil solution have charges ranging from

Figure 4.4 The well-structured subsoil of a Terra Rossa in the Coonawarra region, South Australia. The scale is 10 cm. (White, 2003)

+1 to +3 and also differ in size according to the number of water molecules in their hydration shells.[1] Depending on their hydration energies, cations tend to release these water molecules when attracted close to a clay surface. For example, K^+ has a low hydration energy and readily sheds its water molecules, Ca^{2+} has an intermediate hydration energy and only partially sheds its water molecules, and Na^+ has a high hydration energy and remains hydrated. Thus the force between layers and the distance separating layers varies according to the dominant cations present.

Because cations accumulate in the spaces between clay layers, water molecules try to diffuse into these spaces in response to a concentration gradient between the ambient solution and the interlayers. In so doing, they create a pressure that pushes the layers farther apart. This swelling pressure, which changes with the type of cation and the ionic concentration of the soil solution, can have a disruptive effect on aggregate stability. For example, three times as many exchangeable Na^+ ions than Al^{3+} ions are required for charge balance in a clay crystal, and the Na^+ ions are highly hydrated; thus the tendency for water to diffuse into the crystal, and hence the swelling pressure, is much greater in a Na-clay than an Al-clay.

Aggregate Stability

Apart from an aggregate's size, shape, and color, a key characteristic is its ability to resist "slaking" as it wets up. When water is rapidly absorbed, trapped air exerts a disruptive pressure, augmenting the swelling pressure that develops between the clay layers and around the clay particles themselves. Altogether, these pressures may exceed the forces holding an aggregate together, causing it to collapse or slake. Because of their particular chemical and biological properties, the aggregates shown in figures 4.3 and 4.4 do not slake. A further change that may occur when a dry aggregate wets and slakes is that the swelling pressure is sufficient to push the clay particles so far apart that they disperse and go into suspension. Box 4.3 describes the consequent effects of swelling pressure in more detail. Figure 4.5 shows a simple test for clay dispersion that can be done in the vineyard.

Soil Strength

Soil consistence refers to aggregate strength, which depends on the forces holding an aggregate together. The resilience of a soil's structure is affected by aggregate consistence, which is very dependent on the water content. In the vineyard, aggregate consistence is assessed by the force required to break down a clod or larger aggregate, about 20 mm in diameter, by squeezing between forefinger and thumb or by pressure applied underfoot. The consistence scale ranges from loose (zero)—aggregates

---

[1] A hydration shell is an ordered sheath of water molecules surrounding a cation.

Box 4.3 What Causes Clay to Disperse?

When Ca^{2+} ions are the predominant exchangeable cations, the clay layers within particles, and whole clay particles, come close together in roughly parallel alignment. The reason is that the attractive force between the flat clay surfaces (negatively charged) and the positively charged cations predominates, and the clay is said to be flocculated. However, as Ca^{2+} ions are progressively replaced by Na^+ ions, the weaker negative-to-positive attraction and greater tendency for water molecules to diffuse between the flat surfaces cause the particles to swell and move farther apart. Also, the swelling pressure resulting from the influx of water increases when the soil solution becomes more dilute, as happens when the soil is very wet. The net effect is that clay particles separate to the point where the weakened forces of attraction are overwhelmed and the clay deflocculates or disperses.

The clay suspension shown in the middle jar in figure B4.3.1 came from a creek in inland Queensland. Water in many Australian inland streams remains "cloudy" because of dispersed clay that is eroded from soils containing exchangeable Na^+. The critical amount of Na^+, expressed as a percentage of the

Figure B4.3.1 A sample of dispersed clay (middle jar) and the same clay (right jar) flocculated with 0.1M calcium chloride ($CaCl_2$) solution. Compare the clear supernatant above the flocculated clay with the clear $CaCl_2$ solution in the left jar.

(continued)

Box 4.3 *(continued)*

soil's cation exchange capacity, above which dispersion occurs ranges from 6% to 15%, depending on the type of clay mineral present and the soil solution concentration. Dispersed clay is easily transported in runoff water, which worsens the tendency for soil to erode (see "The Interaction among Cover Crops, Mulches, and Soil Water" in this chapter). Furthermore, as the soil surface dries, the dispersed clay forms a hard crust that inhibits water infiltration and the emergence of young seedlings.

The jar on the right-hand side of figure B4.3.1 shows the same suspension after a small volume of concentrated $CaCl_2$ solution (taken from the jar on the left-hand side) was added and mixed. The dispersed clay has flocculated and settled to the bottom, and the water is now clear. Flocculated clay is a prerequisite for the formation of small stable aggregates, which in turn clump together to form larger aggregates that are stabilized by "cementing agents" (e.g., $CaCO_3$, Fe and Al oxides) and biologically produced gums and mucilage.

The Na–Ca interaction is important for the flocculation of montmorillonite, illite, and vermiculite clays. For kaolinite, flocculation that depends on an attraction between positive charges on the clay edges and negative charges on the flat surfaces is more important. Such flocculation prevails at a pH less than 6 when the edges are positively charged but breaks down as the pH increases and the edges become negatively charged (see "Retention of Nutrients by Clay Minerals and Oxides," chapter 3).

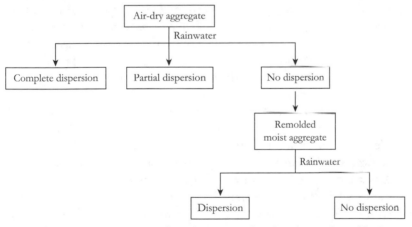

Figure 4.5 A simple dispersion test for soil aggregates, based on the test devised by Emerson (1991), modified by Cass (1999). (Redrawn from White, 2003.)

easily fall apart into constituent particles (e.g., sand)—to rigid (point 7)—aggregates cannot be crushed underfoot. The best condition is weak to firm (points 2 to 3) when clods or large aggregates break down easily into a mix of small aggregates—a friable "tilth." Figure 4.6 shows a sample of clay soil under vines where any surface clods break down easily into firm aggregates 10 to 30 mm in size.

Figure 4.6 Naturally friable tilth in a vineyard soil under straw mulch in the Geelong region, Victoria, Australia.

Additionally, soil in place has a bulk strength that depends not only on its aggregate consistence but also on its bulk density and water content. This relationship is shown in figure 4.7, where the units of strength are those of pressure, commonly, megaPascals (MPa). Box 4.4 outlines how to measure bulk soil strength.

Ideally, strength in the A horizon of a moist soil profile should provide sufficient load-bearing capacity for machinery while permitting vine root exploration of mid-row soil. Measured at the *FC*, the soil strength should be less than 2 MPa and not exceed 3 MPa at the *PWP*. Maintaining the strength of moist subsoils below 2 MPa is also critical for allowing vine roots to explore soil water and nutrient reserves. For example, a survey of 18 drip-irrigated vineyards in southern Australia found that excessive soil strength decreased the effective *AWC* in their subsoils (Murray, 2010). The interaction between soil strength and *AWC*, expressed through the concept of a "non-limiting water range," is illustrated in figure 4.8. When the soil structure deteriorates and bulk density increases, the soil's porosity, especially the proportion of large pores, decreases. As a result, the water content at which aeration becomes limiting decreases, whereas the water content at which soil strength becomes too high increases. The net effect is that the range of water content in which roots can function effectively—the non-limiting water range—is markedly compressed.

A soil of inherently low strength is susceptible to compaction if trafficked by machinery when too wet. Permanent grass swards in the mid-rows are effective in drying the soil and maintaining soil strength in the optimum range. Winter cover crops have a similar effect during what is usually the wettest

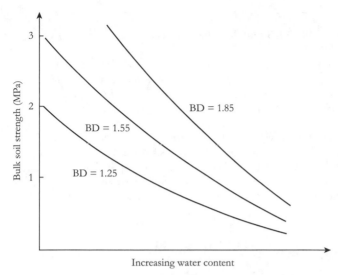

Figure 4.7 Bulk soil strength depends on soil water content and bulk density (BD). (Redrawn from White, 2003.)

period of the year for vineyards. For machinery traveling between the rows, compaction can occur in wheel tracks close to the vines, which can restrict vine root exploration of the mid-rows. However, the effect of compaction is potentially not as great with over-the-row machinery because the wheels travel along the middle of each row. Where mid-rows are repeatedly cultivated, a compacted layer can develop just below cultivation depth. Generally, compaction is more severe in sandy soils than clay loams and clays, but soils with sodic clay subsoils are an exception to this generalization. In Australia, the critical Na content for a sodic clay, expressed as a percentage of the soil's cation exchange capacity, is ≥6% (see box 4.3).

Examples of problems arising from different forms of soil compaction in vineyards are described in "Cultivation and Ripping," chapter 2. However, for high-vigor sites, some wheel-track compaction may be of benefit in restricting the access of vine roots to water and nutrient reserves in the mid-rows.

Water in Soil

Forces Acting on Water

The two most important forces acting on soil water are gravity and suction. The gravitational force simply depends on the height of the water with respect to a reference level, usually the soil surface. Under the influence of gravity, water lying on the surface will seep into the soil or flow downslope.

Box 4.4 Measuring Bulk Soil Strength in the Vineyard

Bulk soil strength in the vineyard is measured by the resistance offered to the entry of a penetrometer, an instrument consisting of a steel rod about 1 m long with a sharp, conical tip and a pressure gauge at the top (figure B4.4.1). As the tip is pushed into the soil, the pressure applied to overcome the soil's resistance is recorded by the gauge. The rod has a graduated scale so that the pressure can be recorded at known intervals down the profile. Readings are most reliable when the rate of entry of the rod is constant. Penetration resistance and hence bulk soil strength vary spatially so that a large number of readings are required to obtain an acceptable average value. Because soil strength depends on wetness, as shown in figure 4.7, measurements should be made at the same water content, usually around *FC*.

Figure B4.4.1
A handheld penetrometer in use. (Photo courtesy of Dr. Jonathan Holland, New South Wales Department of Primary Industries, Australia.)

Penetrometers are valuable for detecting where compacted soil and hard layers occur in the profile and how serious they are. This information allows the most appropriate ameliorative treatment to be determined, whether it be deep ripping, sowing a permanent cover crop, applying gypsum, mulching, or some combination of these.

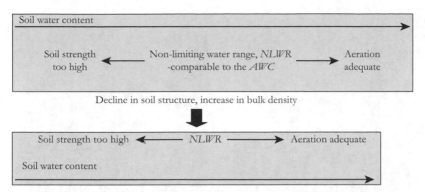

Figure 4.8 A diagram illustrating how the non-limiting water range for vines decreases as subsoil structure deteriorates. Soil water content increases from left to right. (Adapted from Letey, 1985, and Murray, 2010.)

The action of the suction force on water is more complex. As discussed previously, an integral part of a soil's structure is the network of connected pores of different sizes and shapes. The narrower a soil pore, the stronger the surface tension force drawing water into that pore. Hence, water in contact with dry soil is preferentially sucked into the narrowest pores, and if the supply is maintained, water progressively fills larger and larger pores until all the pores are filled and the soil is saturated. The last drop of water to enter the largest pore is held at zero suction.

The combined action of suction and gravity can be illustrated by the behavior of a sponge placed in a shallow dish of water (figure 4.9). Water is sucked into the sponge until it wets to a certain height—this is referred to as capillary rise. The upper part of the sponge does not become wet because the suction drawing water up is canceled out by gravity pulling water down to the "free" water level in the open dish. The net result of these two forces—suction and gravity—determines the hydraulic head of the water. Exactly the same situation occurs in a vineyard soil. In the absence of surface evaporation, the soil water is at equilibrium when the suction force on the water is equal but opposite to the gravitational force at each point in the soil profile. The suction force resulting from surface tension at air–water–solid interfaces is called the matric suction.

At some level, usually at depth in the profile, the soil may be saturated and we find groundwater. The top of this groundwater is called the water table, where the matric suction is zero. This is illustrated in figure 4.10, which shows the hydraulic head profile in a drying soil with a water table at 90 cm depth. Relative to atmospheric pressure, water below the water table has a positive hydrostatic pressure that increases with depth. Hydraulic head differences arise because of differences in groundwater depth and land height, with the result that groundwater flows in the direction of minimizing these head differences.

Figure 4.9 Capillary rise of water. The dyed water is less than 1 mm deep in the glass dish and has risen approximately 10 mm into the sponge in a uniform front.

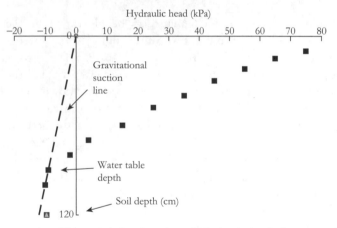

Figure 4.10 The curved plot shows how the hydraulic head of water in a drying profile changes with the matric and gravitational suctions above and below a water table at 90 cm. The gravitational suction is zero at the soil surface; the matric suction is zero at the water table.

This simple picture of soil water must be modified when there are dissolved salts, as occur in saline soils. Salty soil water has an osmotic suction, with the result that the combination of matric and osmotic suction forces is greater than for pure, free water at the same location. Consequently, at the surface of a saline soil, the total force pulling water upward is greater than in a nonsaline soil of the same water content. Osmotic suction causes saline groundwater to rise by capillarity from depths of 1 m in sandy soils and up to 2 m in clay loams, which can create salinity problems for grapevines. The management of soil salinity is discussed later in this chapter in "Drainage, Leaching, and Salinity Control."

Availability of Water to Vines

Soil Infiltration and Wetting

Infiltration starts when water falls on a soil as rain or irrigation. Both matric suction and gravity pull water into the soil, with suction being dominant initially when the soil is dry but negligible when it is thoroughly wet. If water is applied rapidly, as with heavy rainfall, the capacity of the soil to accept it may soon be exceeded, with the result that water collects on the surface and begins to run off. Running water can cause erosion, especially on steep slopes when the soil is bare. Box 4.5 discusses ways of measuring rainfall.

As infiltration proceeds, a zone of wet soil extends downward from the surface. Figure 4.11 is a good example showing a "wetting front" clearly separating wet soil above from drier soil below. Although suction remains important for drawing water into pores at the wetting front, gravity is the dominant force driving flow through the bulk of the wet zone above. Box 4.6 describes how to estimate the depth to which water has penetrated after rain or irrigation.

In clay soils that crack when dry (see figure 4.1), infiltration and wetting do not occur in such an orderly way. Rather water flows rapidly down the cracks so that the soil columns between the cracks wet from the sides as well as the top. This type of flow is called preferential or bypass flow, implying that the water flows down preferred pathways between aggregates more quickly than within aggregates.

Another situation causing irregular infiltration is when the surface soil becomes water repellent (figure 4.12). Water repellence is most noticeable in sandy soils that have become very dry. Organic matter coating the sand grains creates a hydrophobic film, repelling water that may then be "funneled" down old root channels and worm holes. The effect gradually disappears as the soil wets up under persistent rain and can be ameliorated by mixing subsoil clay with the topsoil.

Redistribution of Water in the Soil

Water continues to flow even after infiltration ceases, but flow becomes progressively slower as the largest pores and cracks empty of water, followed by the next largest, and so on. The soil water content attained after two days drainage in a well-structured soil defines the soil's *FC*, as described in "Water Storage" earlier

Box 4.5 How to Measure Rainfall

Rainfall is measured using a rain gauge such as that shown in figure B4.5.1. The gauge should be installed at a standard height (the rim should be 30 cm above the ground in Australia), because wind turbulence causes the amount of rain collected to decrease with the height of the gauge. The gauge should not be overshadowed by vegetation or buildings, nor too exposed to wind. Read the gauge daily at a standard time, usually 9 AM. Rain is measured in millimeters or inches (for conversion, see appendix 1), and 1 mm of rain is equivalent to 1 L per m^2 of ground surface.

Rainfall rate or intensity is measured with a recording gauge that usually has two small buckets of capacity equivalent to 0.1 to 0.2 mm of rain, which tip alternately. An electrical switch sends a pulse to a data logger each time a bucket tips, so rainfall intensity can be measured very accurately in millimeters per minute.

Some advisers in Australia and California calculate "effective" rainfall as a fraction (65%–80%) of measured rainfall. The concept of effective rainfall allows for surface runoff, which is obviously rain that does not infiltrate the soil. However, it should not be used to correct for rainfall intercepted by the canopy and subsequently evaporated, because the energy used in evaporating this water is not available to drive transpiration from the leaves. Thus, although the intercepted water does not reach the soil, the vines' transpiration loss is reduced and the balance between actual evapotranspiration and rainfall does not change.

Winegrowers should use their knowledge of rainfall intensity and vineyard conditions (soil surface condition, mulching, and slope) to decide whether to correct for effective rainfall.

Figure B4.5.1
A standard rain gauge.
(White, 2003)

Figure 4.11 A soil profile showing a clearly defined wetting front. (White, 2006; reprinted with permission of Wiley-Blackwell Publishing Ltd.)

Box 4.6 Estimating the Depth of Soil Wetting

Rainfall or irrigation of 20 mm means that water is deposited on 1 m² of soil surface to a depth of 20 mm. As it infiltrates, this water can only occupy the soil's pore space, which is about 50% by volume (0.5 m³/m³). Thus if all the pore space is initially air-filled, the wetting front will penetrate to a depth x_1, where

$$x_1 = 20 / 0.5 = 40 \text{ mm} \tag{B4.6.1}$$

If there is water in the soil initially (even air-dry soil contains some water), the infiltrating water will penetrate farther. For example, if the initial water content θ is 0.1 m³/m³, the effective pore space to be occupied is 0.4 m³/m³ and the depth of penetration x_2 is

$$x_2 = 20 / 0.4 = 50 \text{ mm} \tag{B4.6.2}$$

Some further penetration of water occurs as the wet soil zone drains to its *FC*. Note also that water does not always penetrate as a uniform wetting front because of preferential flow down worm holes and between aggregates (see "Soil Infiltration and Wetting," this chapter).

Figure 4.12 Separate water drops remain on the hydrophobic surface of a dry sandy soil (left side). The same soil on the right (darker soil) has been kept moist and accepts water readily.

in this chapter. The pores drained of water at *FC* are often called macropores, whereas those that retain water are called micropores.

The matric suction corresponding to *FC* is 10 kPa. However, vine roots can extract water at suctions up to 1500 kPa, when the vines lose turgor and wilt. The *PWP* is reached if vines do not recover turgor at night but remain wilted the following day. The choice of 1500 kPa for *PWP* is arbitrary, because wilting depends on a complex interaction between the variety (also whether on own roots or rootstock), the weather conditions, and how fast water can flow through the soil at large suctions. Tensiometers and gypsum blocks are used to measure soil water suction, as explained in table 4.2. The installation of these instruments is discussed later in "Monitoring Soil Water."

Soil Water Retention and Plant Available Water

As the diameter of pores holding water decreases, there is a proportional increase in matric suction, so that grapevines can extract water more easily from wide pores holding water than narrow pores. The relationship between the amount of water held and its matric suction is called the soil water retention characteristic, which is an important soil property.

Figure 4.13 gives an example of water retention curves for a well-structured clay loam at matric suctions up to 600 kPa for the top 10 cm (A) and for 40 to 60 cm of the subsoil (B). From experience, winegrowers have found that soil water held between 10 and 60 kPa suction is readily available (called readily available

Table 4.2 **Instruments for Measuring Soil Suction**

Type of instrument	Basic construction	Suction range	Installation and comments
Standard tensiometer	Cylindrical porous ceramic cup sealed to a water-filled PVC tube connected to a vacuum gauge (above ground).	0–85 kPa (or centibars, cb)	Can be installed at different depths; reading gives the combined gravitational and matric components of soil suction, so must be corrected for depth to give the matric suction.[a] Purge air bubbles before reading.
"Loggable" tensiometer	Can be a standard tensiometer or a tensiometer with a semi-conductor sensor installed in the cup; suction is measured by a pressure transducer connected to a data logger.	+100 to –160 kPa	As for a standard tensiometer; cups can be made out of different materials to vary the operational range; can read positive pressures (below a watertable); need to be purged and refilled
Gypsum block ("G block")	Cylinder or rectangular block of porous material with embedded gypsum containing two electrodes to which an alternating current is applied; can be logged.	10–200 kPa for "Lite" blocks; 60–600 kPa for "Heavy" blocks	Electrical resistance of the block changes with its water content, in-built calibration of suction-to-water content relationship; not affected by salinity up to 5 dS/m. Correct for depth to obtain matric suction. "Lite" blocks for sandy soils and "Heavy" blocks for clays.

[a] For gravitational suction at zero at the soil surface, matric suction at depth is equal to the tensiometer or gypsum block reading less 1 kPa for every 10 cm of depth (see figure 4.10); positive values indicate a positive pressure.

water, *RAW*), whereas that held between 60 and 400 kPa is less easily extracted (called deficit available water, or *DAW*).

A suction of 400 kPa is called the stress limit, indicating that although vines can extract water held at greater suctions, the rate of supply is not fast enough for them to function optimally. The stress limit changes with soil texture: it is 100 kPa in sandy soils, 200 kPa in loams/clay loams, and 400 kPa in clay soils. These stress limits apply to mature vines. Vines less than three years old should not be subjected to suctions greater than 60 kPa in the mid-root zone.

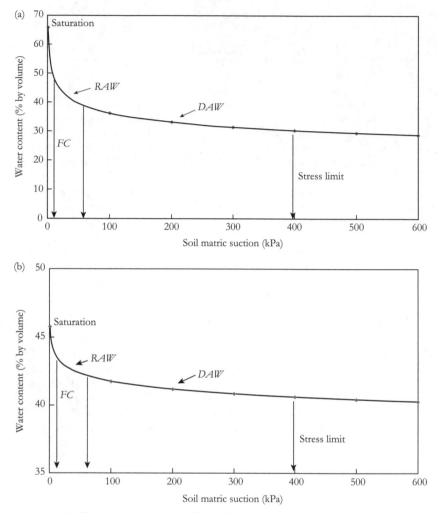

Figure 4.13 Soil water retention curves for a well-structured clay loam. (A) Depth of 0 to 10 cm in the A horizon. (B) Depth of 40 to 60 cm in the B horizon. Readily available water (*RAW*) and deficit available water (*DAW*) are shown for both depths. (Original data courtesy of Dr. Lilanga Balachandra, University of Melbourne, Australia.)

For the top 10 cm of soil in figure 4.13, *RAW* and *DAW* were both 9% by volume ($\theta = 0.09$ m³/m³). From equation B4.2.3, box 4.2, we know that this θ value is equivalent to a depth of water *d* of

$$d = 10 \times 0.09 \times 10 = 9 \text{ mm}$$

for both *RAW* and *DAW*.

In the subsoil at 40 to 60 cm, *RAW* and *DAW* were 1.3% and 1.6% ($\theta = 0.013$ and 0.016 m³/m³), respectively, corresponding to 2.6 and 3.2 mm, respectively, in this depth interval. Thus we see there is generally more *RAW* plus *DAW* in the topsoil than in the subsoil, where the pores are narrower and water is held more tightly.

Values of *RAW* and *DAW* should be calculated for the depth of interest in the soil profile, which may be the depth for irrigation management, or the maximum rooting depth in dry-grown vineyards. Table 4.3 shows results for the clay loam of figure 4.13, calculated using the equation for *d* given above, assuming that the 0 to 10-cm data apply to the whole A horizon (40 cm deep) and the 40 to 60-cm data apply to the whole B horizon (40 to 120-cm deep). The more easily available water (*RAW* plus *DAW*) for 0 to 40-cm (A horizon), 40 to 120-cm (B horizon), and the whole profile (0 to 120-cm) are shown. Also shown is the total plant available water (*PAW*), which includes all the water between *FC* and *PWP* for the relevant soil depths.

Although the relationship between the amount of soil water and its matric suction is mainly influenced by soil structure at small suctions, the effect of texture is more important at large suctions. Because texture is a less changeable property than structure, representative values of *RAW, DAW,* and *PAW* have been developed for soils of different texture. This information, shown in table 4.4, is used for managing the water supply to grapevines under irrigation and also for "Managing Natural Soil Variability in a Vineyard" (see chapter 6).

Table 4.3 Calculated *RAW, DAW,* and *PAW* at Various Depths in the Clay Loam Soil of figure 4.13

Soil depth interval (cm) and texture	*RAW*[a] (mm per depth interval)	*DAW*[a] (mm per depth interval)	(*RAW* + *DAW*)[a] (mm per depth interval)	*PAW*[a] (mm per depth interval)
A horizon				
0–10 (loam)	9	9	18	20
10–40 (clay loam)	27	27	54	60
Total 0–40	36	36	72	80
B horizon				
40–120 (clay)	10	12	23	48
Whole profile				
0–120	46	49	95	128

Note. RAW = readily available water; *DAW* = deficit available water; *PAW* = plant available water.

[a] Calculated using the equation *d* (mm) = 10 × θ × z, where z is measured in centimeters. All numbers rounded to no more than two significant figures.

Table 4.4 The Range of *RAW, DAW,* and *PAW* for Soils of Different Texture (measured in millimeters per centimeter depth of soil)

Soil texture	*RAW*	*DAW*	*RAW + DAW*	*PAW*
Loamy sand	0.55	0.15	0.70	0.86
Sandy loam	0.64	0.28	0.92	1.15
Sandy clay loam	0.71	0.41	1.12	1.43
Clay loam	0.65	0.51	1.16	1.48
Heavy clay	0.41	0.18	0.59	1.20

Note. RAW = readily available water; *DAW* = deficit available water; *PAW* = plant available water. Compiled from Nicholas (2004).

Managing Soil Water with Irrigation

How Vine Growth, Water Stress, and Grape Flavor Are Related

Water stress is a vine's physiological response to a limitation in water supply. A standard rubric of Old World viticulture is that vines must experience some water stress to enhance the sensory sensations of the wines produced. Better wine quality has been linked with lower yields, so that Appellation d'Origine Contrôllée wines in France, for example, normally cannot be made from irrigated vines. This concept has some following in other countries such as Chile and South Africa, where viticulture has been established for a long time. Nevertheless, irrigation is now used to supplement the water supply to vines in many vineyards around the world where summers are hot and dry, because commercial wine production would not be possible without it. Examples are found in the La Mancha region, Spain, the Central Valley of California, the Murray-Darling region, Australia, and the fast-expanding vineyard areas in the Ningxia region of China.

There are several ways of assessing vine water stress (described later). In this section it is discussed in terms of available soil water. When soil water is within the *RAW* range (see figure 4.13), vines are well watered and exhibit actively growing shoot tips and normal stem internode expansion. However, if water is too readily available, especially during the period of rapid shoot growth from flowering to veraison, excess vigor may result, as evidenced by long shoots and a large leaf area and canopy density that cause excessive fruit shading (figure 4.14). Vines display-ing too much canopy relative to the amount of fruit carried are said to be "out of balance." For example, a balanced vine should have a leaf area to fruit weight ratio of 1 to 1.5 m^2 per kg, which provides sufficient leaf area for satisfactory photosyn-thesis and sugar accumulation but not so much that ripening is retarded.

When moderate water stress occurs (soil water content at the bottom of the *DAW* range shown in figure 4.13), shoot growth slows, internodes shorten, and shoot tips may become a dull gray-green. At the hottest part of the day, leaves become flaccid (a sign of temporary wilting) and warm to the touch. With more severe stress (soil water below the *DAW* range and suctions greater than 400 kPa), shoot growth stops and the tips and tendrils may die. Late in the season, water-stressed basal leaves become yellow and may develop necrotic areas toward the edges. Early senescence and leaf fall begin at the base of a shoot and progress toward the tip. Figure 4.15 shows an example of water-stressed vines in late summer in the Napa region, California.

Figure 4.14 Ten-year-old Shiraz vines showing vigorous growth under subsurface drip irrigation in the McLaren Vale region, South Australia.

Figure 4.15 Cabernet Sauvignon vines showing signs of severe water stress (leaf death) in late summer in the Napa region, California. See color insert.

Water stress at flowering can reduce fruit set. From fruit set to veraison, moderate to severe stress reduces berry size through its effect on cell division and enlargement. During ripening, mild water stress enhances the accumulation of soluble solids in the berries by suppressing vegetative growth, but more severe stress decreases berry size through its effect on cell expansion (commonly seen as "berry shrivel"). In this case, sugar accumulation and flavor development are delayed as a result of decreased photosynthesis and premature leaf fall.

In cool, humid climates such as the Burgundy region, France, and humid, maritime climates such as the Médoc region, France, dry-grown vines do not usually experience water stress from flowering to veraison because the supply of soil water is adequate. During the final ripening period, however, a gradual increase in stress enhances the intensity of flavors in the berries so that outstanding vintages are produced in years of dry and hot mid- to late summers. This natural response to a changing water supply seldom occurs in hot inland regions where irrigation is essential to grow grapes. Instead, winegrowers rely on manipulating the amount and timing of irrigation to control vigor and enhance grape quality, as discussed later in "Controlling the Soil Water Deficit through Irrigation."

How a Soil Water Deficit Develops

▪ Evaporation and Transpiration

In winter-rainfall Mediterranean and cool, humid climates, the soil profile under dormant vines is normally at *FC* by the end of winter. During spring, evaporation rates increase as a result of more sunshine and rising temperatures. Evaporation is the process of liquid water conversion to water vapor: it is driven primarily by the absorption of radiant energy from the sun.

A fraction of the solar radiation received is directly reflected, depending on surface reflectance or albedo (literally, its "whiteness"). Some of the short-wave radiation absorbed by the soil and vegetation is reradiated to the atmosphere as long-wave radiation. The greatest part of the remaining "net radiation" (i.e., incoming minus reflected radiation) is dissipated through the evaporation of water (from soil and vines), and the rest is partitioned between heat transmitted to the air above ground and heat penetrating deeper into the soil. Heat transferred to the air aids grape ripening whereas heat transferred into the ground raises the soil temperature. Figure 4.16 shows a summary of these energy transfers.

Water evaporates from moist bare soil at a maximum rate determined by the evaporative demand of the atmosphere until the surface 1 to 2 cm begins to dry. Similarly, after bud burst occurs and leaves expand, the vine transpires water at a rate determined by the air's evaporative demand. The combined effect of these "atmosphere-driven" processes, not limited by soil water supply, produces potential

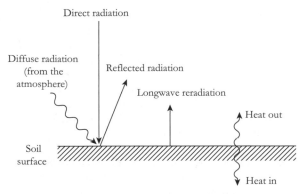

Figure 4.16 Radiation and heat energy balance at a bare soil surface. (Redrawn from White, 2003.)

evapotranspiration (*PET*), the rate of which can be estimated as described in box 4.7. As the surface layer becomes air-dry, soil evaporation slows to virtually zero.

A soil water deficit (*SWD*) develops as water is withdrawn by evapotranspiration (*ET*). Box 4.8 explains how a *SWD* can be estimated. A vine accessing *RAW* normally continues to transpire at or near the maximum rate, and the deeper its roots have penetrated, the more *RAW* can be accessed. However, when this source is depleted, the vine begins to draw on *DAW* and to experience some stress; the leaf stomata start to close and transpiration slows. The greater the stress, the longer the stomata stay shut, which has the unfortunate consequence of limiting photosynthesis and sugar production. During this latter phase, the vine's actual *ET* is considerably less than the *PET*.

Without intervention, the seasonal trend in *SWD* in vineyards under winter rainfall is similar to a sine wave—that is, the deficit increases during summer and then decreases as winter rains refill the soil profile, only for the cycle to be repeated the following season. Figure 4.17 shows an example of this seasonal fluctuation in *SWD*. Note that the size of the *SWD* depends on several factors:

- Soil type and the effective rooting depth of the vines
- Leaf area per hectare (related to the number of vines per hectare and their canopy management)
- Whether mulch and/or cover crops are used
- The balance between rainfall and actual *ET* during the growing period

Controlling the *SWD* within limits can bring benefits through decreased vine vigor and, for red varieties in particular, improved fruit quality. When *SWD* is controlled, there is less need for summer trimming and leaf removal. In addition, control of the amount of irrigation water applied not only conserves this

Box 4.7 Estimating Evapotranspiration

The *PET* of a vegetated soil surface depends on the radiant energy absorbed, the relative humidity of the air, and the wind speed. These are called meteorological or weather variables, the values of which are measured with a weather station (figure B4.7.1). The calculation of *PET* (also called reference *ET*) assumes a reference crop of green grass 12 cm high and completely covering the soil surface, with an unlimited supply of water. Data for calculating *PET* (measured in millimeters per day) are supplied by meteorological services. In Australia, for example, the SILO climate dataset (www.longpaddock.qld.gov.au/silo/about.html) is hosted by the current Department of Science, Information Technology, Innovation and the Arts of the Queensland state government. The weather variables range from rainfall, temperature, solar radiation, and so on to *PET* calculated by five different methods. The data provided are for individual Bureau of Meteorology stations (www.bom.gov.au) or are interpolated on a 5 × 5 km grid for the whole continent.

Figure B4.7.1 Example of a weather station in a vineyard. (White, 2003)

A simple way to measure evaporation is with a shallow tank of water exposed to wind and sun, such as the widely used class A pan shown in figure B4.7.2. The pan is raised above the soil surface to allow air to circulate and so minimize overheating. The daily fall in water level (after allowing for any rain) is measured to give the pan evaporation rate E_p (measured in millimeters per day). Pan evaporation rates are usually about 20% greater than *PET* at the same site.

(continued)

Box 4.7 *(continued)*

Figure B4.7.2 A Class A evaporation pan. The pan diameter is 1.2 m and the depth is 254 mm. The water level is kept 54 mm from the top rim.

Even when there is an adequate supply of soil water, *ET* from a vineyard does not necessarily occur at the potential rate because the canopy may not be fully developed. Research in California has shown that water use by well-watered vines is linearly related to the percentage of the soil surface shaded by the canopy (at midday), achieving a maximum rate at about 75% shading. To account for this effect, *PET* is adjusted with a crop coefficient C_c to give an adjusted *ET* as follows

$$\text{Adjusted } ET = PET \times C_c \tag{B4.7.1}$$

Similarly, if E_p data only are available, the adjusted *ET* is obtained using a crop factor C_f according to the equation

$$\text{Adjusted } ET = E_p \times C_f \tag{B4.7.2}$$

Table B4.7.1 gives an example of crop coefficients and crop factors for two vineyards in southern Australia. Note that C_f values are consistently less than C_c because E_p is normally greater than *PET*. Note also that if deficit irrigation is

(continued)

Box 4.7 *(continued)*

implemented, the coefficients or factors are scaled down during this phase because, when some water stress occurs, the actual *ET* is less than the adjusted *ET*.

Table B4.7.1 **Crop Factors and Crop Coefficients for Vines at Different Growth Stages in a Temperate Climate**

At the growth stage of:	Crop factor C_f^a		Crop coefficient C_c^b	
	No stress	RDI[c]	No stress	RDI[c]
Bud burst[d]	0.2–0.3	0.2–0.3	0.2–0.3	0.2–0.3
Flowering	0.35	0.35	0.45	0.45
Fruit set	0.45	0.25	0.6	0.35
Veraison	0.5	0.3	0.6	0.45
Harvest	0.5	0.5	0.6	0.6
Postharvest	0.25	0.25	0.3	0.3

[a] The crop factors are for healthy vines growing on a sandy loam soil on a single trellis up to 2 m high, with a mown, mid-row cover crop. The canopy coverage increases from 10% of the ground at bud burst to 40% at veraison.
[b] The crop coefficients are for similarly managed vines and canopy cover, except that the mid-rows are bare. Crop coefficients for vines with a cover crop would be slightly larger.
[c] Regulated deficit irrigation (RDI) applied from fruit set to veraison to maintain soil water supply between the no stress and stress limit (see "Soil Water Retention and Plant Available Water," this chapter).
[d] The value at bud burst depends on whether the soil surface is moistened by spring rains (higher value) or not (lower value).
Compiled from López-Urrea et al. (2012), Nicholas (2004), Prichard and Verdegaal (2001), and Goodwin (1995).

Crop factors and crop coefficients may need to be adjusted for local conditions, depending on site potential, vine spacing, presence of a cover crop, and mulching. For relatively small blocks in hot, dry regions, the crop coefficient C_c for a fully irrigated vineyard at large canopy cover (>75%) may be 1 to 1.1, because of the extra energy transferred to the canopy by hot winds blowing from the dry surroundings (called the "oasis effect").

increasingly valuable commodity but also reduces the pumping costs of pressurized irrigation systems.

Controlling the Soil Water Deficit through Irrigation

Regulated deficit irrigation (RDI) offers winegrowers a way to improve water use efficiency and meet their objectives for crop yield and fruit quality. Within these broad objectives, there may be specific objectives depending on the variety (and

Box 4.8 Calculating a Soil Water Deficit

The *SWD* is estimated relative to the soil profile water content at *FC* (sometimes called the "refill point"). Normally, in winter rainfall regions, the soil regains *FC* by bud burst as a result of winter rain, or, when this rain is insufficient, supplementary irrigation may be used. The water content (in millimeters) in the full profile at *FC* is best calculated from the average of θ measurements made toward the end of winter (see equation B4.2.3, box 4.2). Alternatively, *PAW*, estimated from the textures of soil layers as shown in table 4.4, can be used as an estimate of the refill point.

At the refill point, the *SWD* is taken as zero and a deficit develops as the soil water content falls below this value. Above the refill point, the soil holds surplus water that can drain away within one to two days and is not considered to be available to plants. By convention, the *SWD* is plotted as a negative value, and any water held above the refill point is plotted as a positive value (see figure 4.17).

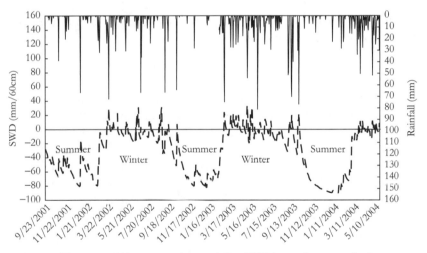

Figure 4.17 The seasonal change in soil water deficit (*SWD*) over three years for Sauvignon Blanc without irrigation in a cool-climate vineyard in Victoria, Australia. Rainfall (solid lines); *SWD* (dashed line). When the *SWD* is positive, drainage can occur. Dates are in U.S. format. (Original data courtesy of Dr. Lilanga Balachandra, University of Melbourne, Australia.)

the scion–rootstock combination) and the style of wine to be made. For example, Shiraz is a vigorous variety often requiring a more severe RDI regime to optimize quality than Merlot, which is more sensitive to water stress. Cabernet Sauvignon is intermediate. Furthermore, rootstocks confer differential tolerance to water stress on the scion varieties, as discussed in "Rootstocks" in chapter 5.

The preferred RDI method uses a threshold deficit, which is the deficit that, when exceeded, signals the need for irrigation. In Australia, RDI is usually applied between fruit set and veraison, when the threshold deficit for irrigating is equated to *RAW* plus some fraction of *DAW*. Take the example given in table 4.3. For the 0 to 40-cm depth (A horizon), the threshold could be set at *RAW* (= 36 mm) plus one half of *DAW* (= 18 mm), giving a total of 54 mm. This amount is 68% of the *PAW* to 40-cm depth. A smaller deficit is advocated from veraison to harvest so that shoot growth is controlled, but sugar accumulation in the fruit is not inhibited. Methods of applying the water are discussed later in "Irrigation Methods."

Sustained deficit irrigation (SDI) is another approach to deficit irrigation that has been applied in inland regions of southeastern Australia, where in drought years irrigation water may be in short supply (see "Managing Soil Water," chapter 6). This approach is analogous to the "volume balance" concept, as practiced in California's Central Valley region. There the threshold deficit is set by the leaf water potential (discussed later), and, when the threshold potential is reached, the vine's weekly water supply is reduced to 60% to 70% of its full demand until after harvest. The vines' full demand is assessed from *PET* data.

Methods for monitoring vine water stress through soil or plant measurements are discussed in the following paragraphs.

Monitoring Soil Water

As explained in table 4.2, one way of monitoring soil water is to measure matric suction with tensiometers or gypsum blocks. The trigger for irrigation to start is set between 60 and 400 kPa matric suction, depending on the soil's texture and the degree of stress to be imposed.

Another way is to measure the volumetric water content θ. For θ measurements, the preferred instruments are those that allow repeated nondestructive sampling, such as capacitance probes (based on frequency domain reflectometry) or probes based on time domain reflectometry (TDR). Box 4.9 describes this equipment in more detail. Neutron probes are generally not used now for health and safety reasons.

The values obtained for all these measurements depend on soil type, depth, and distance from the vine. Because all vineyards show some soil variation, there should be at least one measuring site per soil type with at least two depths of measurement—one in the zone of maximum root density (20 to 30-cm depth) and the other near the base of the "managed" root zone (50 to 60 cm). More sites may be needed in undulating land where soil moisture is likely to vary between the top and bottom of slopes, but a compromise must be made between the cost of more sensors and the value of the extra information obtained. Tensiometers, gypsum blocks, or access tubes for a capacitance probe should be placed in the rows and between vines at a distance of at least 10 to 15 cm (sandy soil) or 20 to 25 cm (clay soil) from an emitter or minisprinkler.

Box 4.9 Instruments for Measuring Soil Water Content

In dry-grown vineyards, soil water content θ should be measured at two to three depths, down to the bottom of the root zone. In irrigated vineyards, the recommended practice is to manage the water content within the top 50 cm of soil, so this defines the depth of interest for monitoring water content.

Capacitance and TDR probes rely on measuring the change in the apparent dielectric constant of soil with a change in water content (water has a very high dielectric constant compared with soil solids and air). A capacitance probe can reside permanently in a narrow-bore PVC tube installed in the soil. A network of such probes installed in a vineyard block can be connected by WiFi to a data logger and readings recorded regularly. Alternatively, a portable instrument can be lowered to different depths in a network of PVC tubes and read directly. Correct installation of the PVC tube is important to avoid air gaps between the tube and surrounding soil; otherwise, false readings are obtained. The annulus of soil that is sensed around the tube wall is small.

For a TDR (sometimes called a Theta probe), an electromagnetic pulse is transmitted down parallel steel wave-guides, up to 30 cm long, inserted into the soil. The time taken for the reflected pulse to return to a receiver is proportional to the soil's apparent dielectric constant and hence varies with the water content. TDR probes may be installed permanently at different depths and logged. Alternatively, portable probes with the pulse generator and receiver built into the probe head can be used for measurements of surface soil water content. The TDR measures an average θ in a narrow cylinder of soil surrounding the wave guides. Modified TDR probes are available to measure bulk EC up to 6 dS/m, as well as water content.

Nicholas (2004), Prichard et al. (2004), and Charlesworth (2000) give further details of these instruments, which are also available on commercial websites.

Box 4.10 describes the calculation of SWD from measurements of soil water content or ET and its use in irrigation scheduling. "Managing Natural Soil Variability in a Vineyard" in chapter 6 discusses irrigation planning to take account of soil variability; the relationship between water stress and wine quality is discussed in "Managing Soil Water" in the same chapter.

Plant Measurements of Water Status

A common plant-based measurement of water status focuses on leaf or stem turgor, which reflects the suction experienced by water within a vine. The suction is expressed as a water potential, a negative variable that is measured as either a stem water potential or a leaf water potential (LWP). For LWP, a young mature leaf in the upper canopy is enclosed briefly in a polythene bag before being plucked and placed in a pressure chamber with the cut end of the petiole

Box 4.10 Use of a Soil Water Deficit to Implement Irrigation Scheduling

If soil monitoring is used, the profile water content is measured at regular intervals during spring through summer (automated systems may provide water content readings on an hourly basis). As indicated in box 4.8, the *SWD* is calculated as the difference between *FC* (in mm) and the profile water content (in mm) on each occasion. Figure 4.17 showed the seasonal trend in *SWD*, calculated in this way, in a nonirrigated block of vines growing on the clay loam soil of table 4.3.

When the accumulated *SWD* reaches a threshold value, which might be set at the *RAW* plus half *DAW* to the depth of interest, the trigger point for irrigation is reached. Figure B4.10.1 shows the effect of RDI, applied from fruit set to veraison, in managing the *SWD* in the same vineyard as shown in figure 4.17 but on a separate block of vines. Because there was little rain from January to March, this period was suitable for applying RDI. The changes in *SWD* and timing of irrigation are shown in the figure.

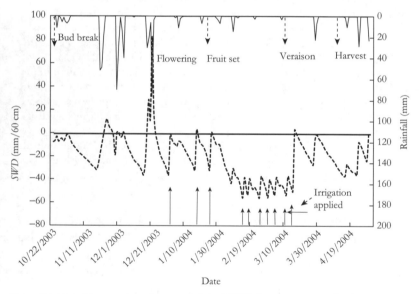

Figure B4.10.1 Changes in soil water deficit (*SWD*) during one year for Sauvignon Blanc vines under regulated deficit irrigation in a cool-climate vineyard. With irrigation, the *SWD* was kept in the readily available water range (0 to 39 mm/60 cm depth) until fruit set, then in the deficit available water range (39 to 58 mm/60 cm depth) up to veraison. The total water applied was 298 mm. Rainfall (solid line); *SWD* (dashed line). Dates are in U.S. format. (Original data courtesy of Dr. Lilanga Balachandra, University of Melbourne, Australia.)

(continued)

Box 4.10 *(continued)*

If *ET* data are used, *FC* at bud burst must be measured or estimated as described earlier. The *SWD* is assumed to be zero then and its subsequent development tracked using weekly rainfall and *ET* data. For example, if the effective rainfall in the first week of bud burst is P_1 millimeters and the actual *ET* is ET_1, the *SWD* at week's end is given by

$$SWD_1 = P_1 - ET_1 \qquad (B4.10.1)$$

Rainfall is measured and actual *ET* estimated, as described in boxes 4.5 and 4.7, respectively. Note that if $P_1 > ET_1$, there is surplus water that can be lost as drainage or runoff; conversely, if $P_1 < ET_1$, a deficit develops. For the second week of bud burst when rainfall is P_2 and *ET* is ET_2, the *SWD* is calculated as

$$SWD_2 = SWD_1 + (P_2 - ET_2) \qquad (B4.10.2)$$

This calculation can be repeated for subsequent weeks and the results accumulated through to harvest. As shown in figure B4.10.1, the trigger point for irrigation occurs when the threshold *SWD* is reached and sufficient water is then applied to keep the *SWD* within the desired range, depending on the stage of growth.

The *ET* method is widely used in California's Central Valley region where the growing season weather is very reliable. Over 120 meteorological stations, distributed around the state as part of the California Irrigation Management Information System (CIMIS; www.cimis.water.ca.gov), collect weather data minute by minute and calculate and store daily values. These "raw" values are transmitted via the Internet to a central computer once a day, where they are quality-checked and values of *PET* calculated for the individual stations. Winegrowers who are registered as CIMIS users can access data relevant to their district and adjust daily *PET* to actual *ET* using a crop coefficient (see box 4.7). Irrigation scheduling is then determined by applying the soil water balance equation (e.g., equation B4.10.1) on a sequential daily basis. The CIMIS website has an example of such a calculation.

Daily *PET* values are available from a similar but less coordinated system of weather stations in southern Australia (see mea.com.au/soil-plants-climate/weather/free-weather-data).

emerging (figure 4.18). The gas pressure just sufficient to force sap from the cut end is assumed to be equal but opposite to the *LWP*. For stem water potential, a leaf close to the trunk is enclosed for at least an hour to prevent transpiration and to allow it to equilibrate with the rest of the vine. The leaf is then plucked and measured as for *LWP*. Stem water potential has the advantage of being independent of the leaf's position on the trunk or main branch.

Figure 4.18 Pressure chamber and accompanying gas cylinder for measuring leaf water potential.

Other techniques to measure vine water status include use of a dendrometer to continuously record small changes in trunk diameter, which enables prompt decisions to be made about when to irrigate. Changes in the rate of trunk growth and diurnal changes in trunk diameter correlate well with changes in *LWP*. Another technique uses infrared thermography to measure the canopy and berry temperature, which increases as the vines experience water stress. However, at present, only *LWP* is used routinely in commercial vineyards, as in California's Central Valley region. Research there has shown that shoot growth starts to slow when the midday *LWP* falls to –0.6 MPa or less, but photosynthesis is not affected until –1 MPa is reached. Vines in the range of –1 to –1.2 MPa are only mildly stressed. The normal threshold to trigger RDI is when *LWP* drops below –1.2 MPa, which is approximately equivalent to supplying 60% of full water demand in the Californian "volume balance" approach.

Comparison of Methods

Table 4.5 provides a summary of the advantages and disadvantages of soil, weather, and plant-based measurements for assessing vine water status and scheduling irrigation.

Partial Root Zone Drying

Partial root zone drying (PRD) is a technique that aims to regulate vine growth while saving irrigation water without sacrificing grape yield and quality. The basic principle of PRD is to keep part of a vine's root system moist while the remainder

Table 4.5 Methods of Assessing Vine Water Status

Method	Advantages	Disadvantages
Weather based		Indirectly related to vine water status; not variety specific
Evaporation pan	Simple and direct daily measurement; applies over a whole vineyard	Birds may interfere; readings corrected for rainfall; a crop factor needed for local conditions
Potential evapotranspiration	Measurement of weather variables can be automated; applies over a whole vineyard	Requires a power source; crop coefficient needed for local conditions
Soil based	Can be variety specific	Indirectly related to vine water status; a site-specific measurement
Tensiometers	Direct measure of soil suction; most sensitive in the *RAW* range; can register positive pressures; those with pressure transducers are loggable	Less suitable in the *DAW* range; water inside tensiometer may freeze; require a soil water retention curve to calculate how much water is required for irrigation
Gypsum blocks	Suitable for full range of suctions for RDI; cheap and loggable	Installation disturbs the soil; require an internal calibration to obtain soil matric suctions; life span of three to five years
Capacitance probe (FDR)	Can measure the full range of θ at all depths; loggable (fixed type) or portable; minimal soil disturbance on installation	Air gaps around access tube affect the reading; fixed type is expensive; readings with a portable probe are time-consuming
"Theta" probe (TDR)	Can measure the full range of θ at all depths; loggable; portable for a surface probe	Installation at depths below the surface disturbs the soil; readings affected by high salinity
Plant based *LWP*	Direct measurement of vine water status	Variable from leaf to leaf and vine to vine; time-of-day dependent; cannot be automated; a site-specific measurement
Stem water potential	Direct measurement of vine water status; more sensitive to water deficit than *LWP*	Cannot be automated; time-of day dependent; labor-intensive
Infrared thermography	Directly correlated with leaf temperature; nondestructive and can be automated	Correlation with leaf temperature is specific to soil, weather, and variety; requires complex calibration

Note. RAW = readily available water; *DAW* = deficit available water; RDI = regulated deficit irrigation; FDR = frequency domain reflectometry; TDR = Time domain reflectometry; *LWP* = leaf water potential.
Compiled from AgNote0294 (2009), Shackel (2006), and Charlesworth (2000).

is allowed to dry out. A hormone (abscisic acid) induced in roots in the dry soil "tricks" the vine into responding, as if it were suffering water stress, by partial closure of its leaf stomata, which in turn decreases transpiration and slows shoot growth. It is recommended that the wet and dry zones be alternated to avoid the possibility of the vine adjusting physiologically by limiting root growth on the dry side and expanding it on the wet side.

Alternation is achieved by having an emitter with its own water supply on each side of individual vines in a row. Modern PRD systems consist of two lines molded together, yet working independently, to allow an alternating cycle of watering. Free-draining soils give the best results because the nonirrigated soil dries quickly and the wet and dry zones are easy to keep separate. In hot, dry weather, the cycle may be as short as three to five days, whereas in mild weather it could be as long as 14 days. Soil water or suction sensors should be installed to monitor soil moisture.

In Australia, PRD potentially works best in hot inland regions where little rain falls during the growing season and vines have to rely on irrigation for most of their water. Although yields may be slightly decreased, berry size is usually decreased too and, for red varieties, the concentration of anthocyanins and phenolics in the juice increased. However, the main benefit of PRD is a saving of water up to 50% of full irrigation, which field experiments in Australia indicate is comparable to the effect of SDI (Chalmers et al., 2005).

Interactions among Cover Crops, Mulches, and Soil Water

Cover Crops and Soil Water

Cover crops growing in the mid-rows have a significant effect on the soil water supply to vines. A winter cover crop takes up soil water and, in drying the soil, improves its strength and trafficability (see "Soil Strength," this chapter). It also protects the soil from erosion by winter rains. Figure 4.19 shows a cover crop of winter cereal in the Western Cape Province region, South Africa.

In drying the soil, a cover crop helps to control excess vigor in vines early in the season, which is important on high-potential sites. However, after an extensive review of cover cropping in Australian vineyards, Proffitt et al. (2013) concluded that for a cover crop to be effective in vigor control, vine roots needed to be exploring the mid-row soil. Where vine roots are confined mainly to the soil in-row because of drip irrigation in dry regions or wheel track compaction, a cover crop has little effect on vine vigor. It also follows that a grower can modify competition between the vines and a cover crop by managing the width of the mid-row sward. A complete sward of a vigorous grass such as

Figure 4.19 A vigorous cover crop of winter cereal in the Western Cape Province region, South Africa.

kikuyu, as shown in figure B2.7.1, chapter 2, provides maximum competition and vigor control.

Permanent grass cover crops decrease the amount of nitrogen available to the vines and lessen the need for herbicides to control weeds. Except on high-potential sites, as in the King Valley region, Victoria, Australia, for example, permanent cover crops are not recommended for dry-grown vineyards, or those receiving minimal irrigation, because soil water needs to be conserved during summer. For this reason, winter cover crops, which in spring are mown or cultivated into the soil, are favored in many vineyards. Keeping the mid-rows well mown or free of vegetation in spring and early summer reduces *ET*, often improves crop yields, and reduces the risk of late frost damage to the vines.

"Cover Crops and Mulches" in chapter 5 provides further details on cover crop species and the effect of cover crops on soil biology.

Mulches, Soil Water, and Temperature

When a cover crop is mown in spring, the mowings can be thrown sideways to form a mulch under the vines. Other organic forms of mulch include compost, bark chips, and cereal straw (figure 4.20). About 25 m³ of mulch (approximately

Figure 4.20 A straw mulch in vine rows in the Barossa Valley region, South Australia.

25 t of dry matter) is required for an under vine strip 50 cm wide and 5 cm deep at 2500 vines per hectare.

Mulches have several beneficial effects, including

- Shading the soil, thereby reducing soil evaporation
- Suppressing weeds, thereby reducing the need for herbicides
- Preventing raindrops from impacting directly on the soil surface, thereby maintaining soil structure and water infiltration
- Encouraging biological activity, especially of earthworms, through the added organic matter and by keeping the soil moist

The effect of mulch on soil evaporation is more important while the vines are dormant and before the canopy begins to develop in spring and early summer. Also, the row spacing is an important factor because this affects the degree of shading of the soil by the canopy. When the proportion of shaded ground is less than 10%, soil evaporation may comprise as much as 70% of total vineyard *ET*. This is the period when mulches can have a significant effect in reducing wasteful losses of water through soil evaporation. However, as the canopy develops to shade 40% or more of the ground, soil evaporation is a much smaller fraction of total *ET*, especially for drip-irrigated vineyards when it can be as low as 10 to 20%. Nevertheless, in furrow or flood irrigated vineyards even with mature canopies, soil evaporation has been estimated to be as much as 41% (furrow) to 77% (flood) of *ET* (Kool et al., 2013).

Mulches affect soil temperature. Because mulches have a higher albedo than bare soil, the energy absorbed during the day, and hence soil warming, is less under this type of cover. The fact that soil stays moister under mulch also influences its temperature, because a wet soil requires more heat energy than dry soil for its temperature to rise one degree. Soil wetness is therefore important in determining when root growth starts in spring. Vine roots start to grow around 6ºC and have a temperature optimum close to 30ºC.

Surface albedo is also affected by color, with black and dark-red surfaces reflecting less radiation and therefore warming more quickly than light-colored surfaces. However, at night, dark surfaces radiate energy faster than light-colored surfaces, adding to the loss of heat by convection. Bare soil surfaces with no vegetative cover lose heat rapidly at night, especially if there is no low cloud cover. In the Châteauneuf-du-Pape appellation of the southern Rhone Valley, France, large quartzite stones (called *galets*) deposited by the river over millennia act as "heat sinks" during the day and reradiate this heat at night (figure 4.21). The vine canopy traps some of this heat with the result that the surface soil temperature and near-surface air temperature remain higher than otherwise expected. This is an important factor in decreasing frost damage to green tissues.

Organic mulches, such as straw or bark pieces, have a low thermal conductivity and slow the transfer of heat into and out of the soil. Soil under mulch therefore remains cooler during the day and warmer at night than bare soil, resulting in

Figure 4.21 A vineyard with heat-absorbing *galets* to the east of Châteauneuf-du-Pape appellation in the Rhone Valley region, France.

a decreased diurnal change in soil temperature. Lower daytime soil temperatures decrease the evaporation rate and favor earthworm activity.

The distinction between the function of mulches and composts is discussed in "Cover Crops and Mulches" and "Manures and Composts" in chapter 5.

Drainage, Leaching, and Salinity Control

What Causes Poor Drainage?

The best vineyard soils are invariably freely drained. They are often deep soils (e.g., see figures 1.10 and 1.11), but shallow soils on limestone can also be free draining, provided the limestone is porous (as in the Coonawarra region, South Australia) or fractured and fissured (as in the Côte d'Or region and St. Emilion appellation, France).

An impermeable subsurface horizon, which is often a poorly structured and dense B horizon, impedes drainage. The sodic clay subsoils of some duplex soils in southern Australia are an example, as are some of the duplex soils in the Western Cape Province region, South Africa, and the Piedmont region of eastern United States. In winter, when rainfall exceeds ET for long periods, water accumulates at the top of the B horizon to form a "perched" water table (see figure B1.3.1). The soil above the B horizon becomes waterlogged. Because the solubility of O_2 in water is low, O_2 diffusion through the water-filled pore space is very slow compared with its diffusion in air. Consequently, roots and soil microorganisms are starved of O_2 and cannot respire normally: the soil becomes anaerobic. Although some microorganisms, such as those responsible for denitrification (the reduction of NO_3^- ions), can switch from aerobic to anaerobic respiration and continue to function, vine roots cannot function under such conditions, and the plant's nutrient uptake and growth suffer. Box 1.3, chapter 1, discusses other biochemical redox reactions occurring in waterlogged soils.

Although well-drained soil should not develop a perched water table, the rise of groundwater to within 1 to 2 m of the soil surface (see "Forces Acting on Water," this chapter) can create problems. Because water is rising from deep in the profile where there is little organic matter, the carbon substrates required for anaerobic microbial respiration are in short supply, so undesirable redox reactions are less problematic. However, too much water supplied to the deeper roots encourages excess vigor. Figure 4.22 shows a deep, sandy soil at the bottom of a slope in the St. Emilion appellation where this problem occurs—similarly in some of the heavier textured soils of the Pomerol appellation, Bordeaux region, France. Furthermore, if the groundwater is saline or salts have accumulated in the soil as a result of poor irrigation management, capillary rise of salts can seriously affect vine health, crop yield, and fruit quality.

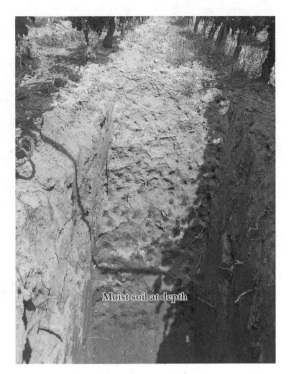

Figure 4.22 A deep sandy loam soil under vines in the St. Emilion appellation, Bordeaux region, France. Note the deep rooting and darker moist soil in the lower profile.

Moist soil at depth

How to Improve Soil Drainage

Poor drainage that results in waterlogging can be caused by compaction and poor soil structure. Box 2.6, chapter 2, discusses deep ripping to break up compacted soil when a vineyard is established and methods of alleviating compaction in the mid-rows of established vineyards. Apart from mechanical intervention, soil structure and hence drainage can be improved by applying amendments such as gypsum or by growing a cover crop. Grasses and cereals are most effective because of their fibrous root systems, but the deep taproots of species such as chicory are also effective. Mulches and composts also improve structure through the addition of organic matter that encourages earthworms and stimulates soil biological activity.

Mounding or hilling soil under vine increases topsoil depth, thereby potentially increasing the rooting volume. As shown in figure 4.23, mounds also help to shed surface water more rapidly, especially on slopes. Subsurface drainage may be needed to control waterlogging in flat land. In long-established regions, porous clay pipes (called tiles) were laid in rows at a depth between 1 and 2 m to remove excess water and control the water table. For new vineyards, perforated PVC pipe ("aggie" pipe) can be laid by a laser-guided machine on a precise gradient without trenches needing to be excavated. The pipe spacing varies between

Figure 4.23 Steep mounding of a granite-derived soil in a new vineyard in the Western Cape Province region, South Africa.

10 and 40 m, depending on the depth at which the pipes are laid and the soil's hydraulic conductivity. Subsoil drainage with pipes has been used, for example, in Pomerol, France, the Tuscany region, Italy, and the Carneros subregion of the Napa Valley, California. Although the installation of subsoil drainage is expensive, it is often worth the investment when amortized over the commercial life of a vineyard.

Subsoil drainage is sometimes necessary in irrigated vineyards of the Riverina region in Australia to prevent saline groundwater rising into the root zone. In some cases, groundwater pumping and discharge into evaporation basins may also be required.

Leaching, Salinity, and Sodicity Control

Irrigation supplies both water and salts to vines. Because most of the applied water is transpired by the vines and only some of the salts absorbed, the salt concentration in the soil gradually increases as it passes through repeated wetting and drying cycles. This salt increase can be monitored by measuring *EC* at the bottom of the root zone, preferably by obtaining a sample of the soil solution with a suction sampler, such as the SoluSAMPLER (figure 4.24). Significant salt accumulation occurs even with low *EC* irrigation water (<0.8 dS/m), but the problem is worse

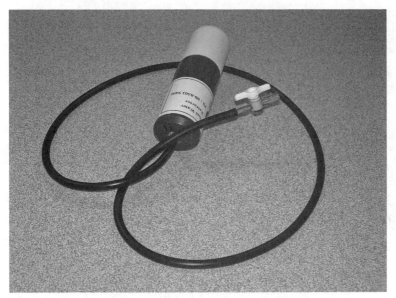

Figure 4.24 A SoluSAMPLER™ is a porous ceramic cylinder sealed to PVC tubing with a vacuum tap inline. Suctions up to 60 to 70 kPa are applied with a syringe attached to the open tube. (Photo courtesy of Dr. Tapas Biswas, formerly of the South Australian Research and Development Institute, Adelaide, South Australia.)

when water of higher salinity is used. Examples occur with some of the groundwater used in the Barossa Valley region, South Australia, the Darling River water in the Murray-Darling region, Australia, and the Tajo River water in the Castilla la Mancha region, Spain. Sometimes a white salt encrustation can be seen on the soil surface, as shown in figure 4.25.

Increasing soil salinity predisposes to a soil becoming sodic, as defined in box 2.4, chapter 2. Saline-sodic soils are potentially unstable when rainfall or irrigation water of low salinity gradually leaches out the resident salts. The lowered salt concentration, coupled with a high exchangeable sodium percentage, causes clay to disperse with a consequent destabilization of soil structure, as explained in box 4.3. The first sign of structural damage is a decrease in the soil's permeability (i.e., its hydraulic conductivity). This effect is most noticeable in a clay subsoil in which dispersed clay can block soil pores, reducing permeability to a very low value. Waterlogging is likely to ensue.

Sodic soils should be treated with gypsum to replace Na^+ ions with Ca^{2+} and so promote clay flocculation and stable aggregation. Topsoil sodicity is easy to treat with surface applications of gypsum (1 to 2 t/ha/year) for two to three years, but subsoil sodicity may require deep ripping and concurrent gypsum application (see "Cultivation and Ripping," chapter 2).

Figure 4.25 A white salt encrustation under a dripper line in a vineyard in Castilla la Mancha region, Spain.

Salt accumulation is counteracted by controlled leaching to maintain a salt balance. As discussed in box 4.11, this involves calculating a leaching requirement (*LR*). The *LR* can be met through regular irrigation scheduling. However, with watering strategies such as RDI, PRD, or SDI, deep drainage during the irrigation periods is minimized. Leaching of accumulated salts then needs to occur during the period of winter dormancy, either through rainfall or a deliberate blanket irrigation with good quality water.

It is important to distinguish between the *LR* and leaching fraction (*LF*), the latter being the actual fraction of applied water that passes through the root zone. Ideally *LF* should equal *LR*, but because of variability in water application and infiltration, *LF* is usually greater than *LR*, which means more water drains through the root zone in some parts of the vineyard than is necessary for salinity control. This occurs more commonly with flood and furrow irrigation because of poor control over the amounts of water applied (see "Irrigation Methods" in this chapter).

Overall Salinity Management

Good salinity management must consider not only a soil's salt balance but also the offsite effects of salt. Unnecessary leaching of salts into drainage water creates salinity problems for water users downstream. "Best-practice" irrigation must therefore focus on the dual objective of minimizing saline return flows to streams and avoiding salt buildup in the soil and in groundwater immediately below the soil.

Box 4.11 Salinity Control by Leaching

To maintain salt balance in the soil, the input and output of salts must be equal. For an irrigated vineyard, accession of salts in rainfall and from fertilizer is approximately canceled out by the salts removed in fruit (and possibly prunings) and precipitation of insoluble compounds, so the salt balance simplifies to

Salts from irrigation water = Salts in drainage below the root zone   (B4.11.1)

Because EC is a surrogate measure of salt concentration, and volume per unit area is equivalent to a depth d of water (in millimeters), the salt balance can be written as

$$EC_{iw}d_{iw} = EC_{dw}d_{dw}$$ (B4.11.2)

where the subscripts iw and dw refer to irrigation and drainage water, respectively. By rearranging this equation, the leaching requirement (LR) can be equated to a ratio of EC values; that is,

$$LR = d_{dw} / d_{iw} = EC_{iw} / EC_{dw}$$ (B4.11.3)

Thus LR is defined as the fraction of irrigation water (conductivity EC_{iw}) that must drain out of the root zone to maintain the EC in that zone (EC_{dw}) at or below a critical value. Experiments indicate the critical saturation extract EC (EC_e) for *Vitus vinifera* is 1.8 dS/m (see "Soil Testing for Salinity," chapter 3), which is approximately equal to an EC_{dw} of 3.6 dS/m. Hence, for irrigation water of 0.8 dS/m, we can calculate the LR as

$$LR = \frac{0.8}{3.6} = 0.22$$ (B4.11.4)

For vines on a salt-tolerant rootstock, the LR would be smaller because the critical EC_e is higher. Conversely, for the same LR, water of higher EC_{iw} could be used.

If one assumes no effective winter rainfall, this calculation shows that about one-fifth of the irrigation water should drain below the root zone to avoid salt accumulation. In this case, for a vineyard water requirement of, say, 300 mm during the growing season, the total amount of irrigation required to satisfy the vines and to provide an LR would be

$$d_{iw} = \frac{300}{1 - 0.22} - 385 \text{ mm}$$ (B4.11.5)

The additional 85 mm of drainage would normally be provided by winter rainfall, so only when winter rainfall is too low would additional irrigation be required during winter to provide the LR. The value of EC_{dw} should be monitored over several seasons to check that leaching is effective in maintaining a salt balance.

Best-practice salinity management under irrigation involves

- Keeping the *LR* as low as possible (≤0.1) by using low-*EC* water wherever available. This may require blending good-quality water with high-*EC* water.
- Using salt-tolerant rootstocks when appropriate (see "Rootstocks," chapter 5)
- Matching the application of water to the vines' demand as assessed visually, from *ET* data, or by monitoring soil water status
- Monitoring drainage so that *LF* is as close as possible to *LR* (see "Monitoring Deep Drainage" in this chapter)

Irrigation Methods

Irrigation methods for vineyards fall into two broad categories:

1. Microirrigation using drippers, microjets, or minisprinklers
2. Macroirrigation using overhead sprinklers, flood, or furrow systems

Topography and soil type influence the choice of system, as well as the availability and price of water, the capital costs of installation, and the skill of vineyard staff.

Microirrigation systems are all under-canopy methods of applying water. They offer the best means of accurately applying predetermined amounts at known rates, targeted to specific areas with minimum losses. Hence, the efficiency of water use from pump to target area can be as high as 80 to 90% for well-managed systems. Although easily automated and controllable by wireless signals, these systems are the most capital-intensive.

Macroirrigation systems are the least efficient, ranging from about 60% for fixed overhead sprinklers to less than 50% for furrow and flood systems. The latter tend to be found in older, established vineyards because of their low installation cost. However, they have become less viable with the increasing cost of water (resulting from scarcity) and because they are time-consuming to operate (and hence have high labor costs).

Microirrigation Systems

Drip Systems

Drip systems consist of emitters (drippers) spaced at regular intervals along flexible PVC pipes, usually suspended 0.3 to 0.5 m above the soil under the vines. However, in an effort to save water, subsurface drippers are also being installed because soil evaporation rates are much lower than with surface applications. Emitter spacing and discharge rate depend on the vine spacing and soil texture, because the aim, particularly in dry regions, is to have a continuous band of moist soil in the rows. Specialist texts such as Nicholas (2004) and Pritchard et al. (2004) give details of the types of dripper available and installation methods.

The wetting pattern under an emitter changes with soil texture. The suction force pulling water sideways and downward is small in sand and sandy loams because of their large pores, and gravity predominates (figure 4.26A). In such soils the emitters can be as close as 0.5 m and, given the high infiltration rate, the discharge rate can be as much as 4 L/hour. As the texture increases to clay

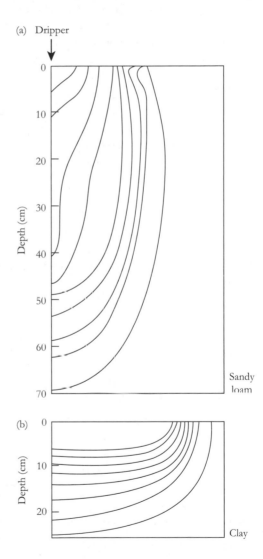

Figure 4.26 (A) Pattern of wetting under a dripper in sandy soil of high permeability. (B) Pattern of wetting under a dripper in clay soil of low permeability. (Redrawn from White, 2006.)

loam or clay, the suction pulling water sideways increases relative to the constant force of gravity so that the wetted zone is more like a flattened onion bulb (figure figure 4.26B). Also, because infiltration rates are slow, the preferred emitter discharge rate is 0.8 to 2 L/hour and the emitters can be placed farther apart.

Table 4.6 shows the amount of water needed to deliver a specified RDI, depending on soil texture and emitter spacing. In practice, if winegrowers are advised to control the water content of the top 0.5 m of soil only, pulsed application of water (30 minutes on and 30 minutes off) may be necessary to prevent deeper wetting in sandy profiles.

Depending on soil depth and permeability, subsurface drip irrigation (SSDI) lines are placed 0.20 to 0.25 m below ground along a vine row (figure 4.27). This method is favored in clay soils where suction is the dominant force drawing water through the soil and drainage below the root zone is minimal; it is not as suitable for sandy soils in which gravitational movement of water is much greater. Because soil evaporation tends to concentrate dissolved salts, SSDI is a good option when reclaimed water is used (see "Reclaimed Water," chapter 2).

Directly pumped river water may contain colloidal sediment that can cause blockages, and some groundwaters may contain sufficiently high concentrations of soluble Fe, which, on exposure to air, precipitates as an insoluble Fe hydroxide (ferrihydrite). In these cases, valves installed in individual drip lines can be monitored to check for blockages. The problem of Fe precipitates can be alleviated by passing through the delivery line a solution of commercially available humic acid, which chelates and dissolves the Fe compound (see "Availability of the Micronutrients and Trace Elements," chapter 3).

Table 4.7 summarizes the advantages and disadvantages of drip irrigation relative to other systems.

Table 4.6 **Examples of Water Requirements for One Application of RDI on Two Soils of Different Texture**

Soil texture and depth	Vine spacing (m)	Irrigation	RDI[a]	Total water required (kL/ha)[b]	Water required if only one third of the area is watered (kL/ha)
Loamy sand, 0.6 m	1.5 × 2.5	4 L/hr emitters at 0.6 m spacing (4444 per hectare)	38	375	125
Clay loam, 0.6 m	1.5 × 3	2 L/hr at 1 m spacing (2222 per hectare)	54	543	181

Note. RDI = regulated deficit irrigation.

[a] Assume the RDI water required is equal to readily available water plus one half of the deficit available water, derived from table 4.4; figures rounded to the nearest mm.

[b] 1 mm = 1 L/m^2 and 1 ha = 10,000 m^2.

Figure 4.27 A subsurface dripline in a loamy soil in a vineyard in the Swan Hill region, Victoria, Australia.

Microjets and Minisprinklers

Microjets, minisprinklers, and under-canopy impactsprinklers deliver water at much higher rates than inline drippers and wet a larger proportion of the ground area. Thus they do not save as much water as drip systems, but they do give better water distribution on sandy soils, thereby allowing a cover crop to be grown in dry regions. As with SSDI, valves in the delivery lines may be necessary to monitor for blockages when filtered water or water with a low concentration of suspended solids is not available. Nicholas (2004) and Prichard et al. (2004) give further details.

Macroirrigation Systems

Overhead Sprinklers

Because overhead sprinklers wet all the ground area, water in excess of plant needs may be applied to parts of the vineyard, leading to drainage losses. Evaporative loss during water application can also be high, especially on windy days, and dissolved salts can cause scorch as water intercepted by leaf surfaces evaporates (figure 4.28).

Table 4.7 **Advantages and Disadvantages of Drip Irrigation**

Advantages	Disadvantages
Water applied to individual vines can be closely controlled; particularly suited to regulated deficit irrigation and partial root-zone drying. With frequent application, soil water content (and suction) is maintained in a narrow range.	Suspended sediment, chemical precipitates, or microbial growths may cause blockage of emitters, as can root invasion for SSDI; frequent flushing helps to prevent this. Vacuum relief valves should be installed at the head of the system and at high points along the laterals.
Little loss by soil evaporation and runoff (especially for SSDI), because area of wet soil is small and rates of application are low; no foliar salt damage as is possible with overhead sprinklers.	Salt accumulates at the edge of the wetted zone, especially if there is insufficient winter rain to leach salts (more serious with water >0.8 dS/m).
Low drainage losses (hence low loss of nutrients and potential contaminants), especially if drainage is monitored.	Rodents, other mammals, and birds seeking water can damage the soft, flexible pipes (not a problem for SSDI).
Targeted fertilizer application by fertigation (see "Nitrogen Cycling," chapter 3).	Less effective control of vine microclimate than with overhead sprinklers, which can be used to avoid frost damage to vines at critical times.
Allows use of water of greater salinity than with overhead sprinklers or flood/furrow systems (where irrigation is less frequent and the soil dries between applications).	More expensive to install and maintain than gravity-flow flood and furrow systems.
Wastewater can be used because chance of pathogens contaminating foliage or fruit is minimal (none for SSDI).	Wastewater may need to be treated to a high standard to prevent excessive bacterial growth and blockages.
Low operating pressures (100–200 kPa) mean low pumping costs; restricted weed growth, especially with SSDI, and no constraints on access because of wet soil.	
Effective on marginal soils and in difficult topography (stony soils, steep slopes)	

Note. SSDI = subsurface drip irrigation.

In frost-prone areas, overhead sprinklers can be used to raise humidity in the vineyard and provide a water film on leaf surfaces when a frost is likely. The high heat capacity of water and the latent heat released as it freezes help to prevent leaf temperatures falling below 0°C. Figure 4.29 shows an example of overhead sprinklers in operation in a vineyard in the high-altitude Rueda district of Castilla y León region, Spain.

Figure 4.28 Salt scorch on leaves in a vineyard in the Murray-Darling region, New South Wales, Australia. (Photo courtesy of Dr. Tapas Biswas, formerly of the South Australian Research and Development Institute, Adelaide, South Australia.)

Figure 4.29 Overhead sprinklers in action in a Sauvignon Blanc vineyard in the Rueda district of Castilla y León region, Spain.

Flood and Furrow Irrigation

For flood irrigation, the mid-row area across to the mounded vine rows on each side comprises an irrigation bay. Water flowing from a head channel into the bay covers the entire mid-row. Sufficient water must be applied for soil at the far end of the bay to wet to a desired depth, which usually means that vines at the head of the bay receive too much water. This uneven distribution is worse in long bays on permeable soils—a problem compounded further by the presence of slight rises and hollows where water ponds and infiltration is greater. Laser-guided grading of the land when the vineyard is established minimizes this problem.

In furrow irrigation, water is diverted from a head ditch into shallow channels, either on one or both sides of a vine row (figure 4.30A), or sometimes into a single median channel (figure 4.30B). Although furrow irrigation suffers from the same disadvantages as flood, both have the advantage of low capital costs and permit the use of water that would be problematic for drip irrigation because of suspended solids.

Furrow and flood irrigation are wasteful of water compared with drippers and under vine sprinklers and are being phased out in most irrigated vineyards around the world.

Monitoring Deep Drainage

Although capacitance or TDR probes can be used to monitor changes in soil water content at depth, because water moves in response to a gradient in suction, gravity, and osmotic forces combined (not a gradient in water content), drainage can occur when changes in water content are small and undetectable by these probes.

Tensiometers placed at the bottom of the root zone are best for detecting deep drainage. As shown in table 4.2, the matric suction at the depth of the tensiometer bulb is equal to the tensiometer reading (measured in kPa) less the placement depth in decimeters (10-cm intervals). A matric suction in the range 0 to 1.0 kPa indicates that the soil is very nearly saturated and drainage is occurring.

The Full Stop Wetting Front detector is another instrument capable of detecting the movement of a wetting front at depth (see www.fullstop.com.au). The Full Stop detector is installed at two depths—usually at 0.3 and 0.5 m—under irrigated vines. A float rises when the matric suction at the funnel rim falls to 2 kPa or less and about 20 mL of drainage water has collected in the reservoir at the base of the detector. Thus not only can the arrival of a wetting front be detected, but a water sample can also be collected for analysis. As with the SoluSAMPLER (see figure 4.24), NO_3^- concentrations can be measured to check on any leaching loss that might lead to groundwater contamination. Also, measurement of EC_{dw} (see box 4.11) at successive times allows an estimate to be made of the amount of drainage, provided that the depth of applied water and its EC are known.

Figure 4.30 (A) Furrow irrigation of Sauvignon Blanc vines on a gravelly soil in the Rueda district of Castilla y León region, Spain.
(B) Single-furrow irrigation between rows of 90-year-old Zinfandel vines in the Lodi District, California.

Summary Points

1. Grapevines are vigorous perennial plants that are potentially deep rooted, depending on the depth of soil and its structure. Soil structure is an expression of how particles of clay, silt, and sand combine to form aggregates and of the nature of the pore spaces, or porosity, created within and between aggregates. Aggregate type and stability, aeration, drainage, water storage, and soil strength are all important factors affecting vine growth that depend on a soil's structure.

2. A predominance of exchangeable Ca^{2+} ions on the clay particles ensures soil aggregates are stable and able to resist the disruptive forces caused by raindrop impact and cultivation. Aggregate stability is enhanced in topsoils by well-humified organic matter, which interacts with the clay particles, and by gums and mucilages secreted by plant roots, bacteria, and fungi. Aggregate stability in subsoils depends more on the presence of Fe and Al oxides and is much less reliant on organic compounds.

3. Aggregation is potentially unstable in sodic soils in which exchangeable Na^+ ions occupy more than 6% of the soil's cation exchange capacity (*CEC*). A combination of too much exchangeable Na^+ and a dilute concentration of salts in the soil solution predisposes clay to disperse or deflocculate and disrupt the structure of small aggregates. Gypsum is recommended for amelioration of sodic soils, and deep ripping may be required to treat subsoil sodicity with gypsum.

4. A soil's pore space is occupied by water and air. When fully wet, the soil is saturated, and, as it drains, air gradually displaces water from the widest pores, followed by increasingly narrow pores. Normally, within two days a soil drains to its upper limit of available water, which is called the field capacity (*FC*) or refill point. The lower limit of available water—the permanent wilting point (*PWP*)—is reached when the vines can no longer extract water at a rate that matches their transpirational demand (they wilt and do not recover turgor at night). The water held between these two limits is the available water capacity (*AWC*), which is a function of both a soil's structure and texture, and is usually expressed in mm water per meter depth of soil.

5. Suction forces and gravity drive water movement into and through a soil. The suction arising from surface tension forces is called matric suction; the concentration of dissolved salts determines the osmotic suction. Gravity is the dominant force when the soil is wet, especially when there are large cracks in which water can flow rapidly downward. In a nonsaline soil, the hydraulic head is the sum of the matric and gravitational components of the soil water suction.

6. As water drains from a saturated soil, matric suction forces acting on the remaining water gradually increase. A soil water retention curve, unique to each soil, describes the change in soil water content as the matric suction increases. At the *FC*, the matric suction is normally 10 kPa, and this increases to 1500 kPa at the *PWP*. The water held between 10 kPa and 60 kPa is considered to be readily available water (*RAW*)—vines do not suffer any stress in this range; that held between 60 kPa and 400 kPa is deficit available water (*DAW*), because in this range vines suffer moderate water stress, with symptoms observable in the vineyard.

7. The fraction of a soil's volume that is drained at the *FC* defines its air-filled porosity, which should be at least 10% for adequate gas exchange (aeration). Anaerobic conditions that are unfavorable to root function are likely to occur when the air-filled porosity is <10%.

8. Soil compaction, which is obvious in sodic subsoils when dry, causes an excessive increase in soil strength. Soil strength, measured with a penetrometer, should ideally be <2 MPa when the soil is at its *FC*. Such a condition provides a sufficiently firm base for machinery, promotes good drainage, and allows easy root penetration.

9. Vineyard soils in winter rainfall regions should normally regain *FC* by the end of winter. After bud burst, evaporation from the soil surface and transpiration through the vines (together evapotranspiration, or *ET*) create a soil water deficit (*SWD*). The *SWD* increases as vines first extract *RAW* and then *DAW*. For the same depth of soil, the sum of *RAW* and *DAW* relative to *AWC* is greatest for sandy clay loams and clay loams and least for heavy clays. The potential plant available water (*PAW*) is the product of *AWC* and the soil depth.

10. In dry-grown vineyards, the supply of *RAW* plus *DAW* depends on the balance between rainfall and *ET*, soil texture, and the depth of rooting. In irrigated vineyards, depth of rooting is less important because irrigation is used to control the water content of the top 50 cm of soil only. In both cases, moderate water stress from fruit set to veraison enhances fruit quality and the flavor and aromas of the wines produced. In irrigated vineyards, regulated deficit irrigation (RDI) is used to control *SWD* in a desired range and also saves water. Partial root zone drying (PRD) under irrigation can also be used to save water and enhance fruit flavors without serious yield loss.

11. Soil water monitoring by means of tensiometers, gypsum blocks, capacitance probes, or time domain reflectometry can be used to schedule irrigations and indicate the amount of water to be applied. Vine water stress is assessed directly by measuring a leaf or stem water potential. Soil water status can also be estimated from measurements

of rainfall and potential *ET* (*PET*), provided that a crop coefficient is used to adjust the *PET* according to the vineyard's stage of growth and canopy cover.

12. Both RDI and PRD are most effectively practiced with drip irrigation, which can be delivered above ground or below (by subsurface drip irrigation) through suitable emitters. Other methods include microjets, minisprinklers, overhead sprinklers, flood, and furrow irrigation. The last two are the least efficient in terms of water delivery but the cheapest to install and operate.

13. Mid-row cover crops and under vine mulch moderate soil temperatures and affect soil water storage and biological activity. Permanent grass swards compete with vines for soil water and on high-potential sites in higher rainfall regions can help to control excess vigor. They also protect soil from erosion.

14. Drainage is determined not only by land slope but also by soil texture and structure, especially of the subsoil. Poor drainage can lead to waterlogging, with consequentially undesirable effects on root growth and soil microbial activity. Drainage can be improved through the installation of perforated pipes at depth, provided the water collected can be moved offsite. Pumping may be necessary to keep groundwater below 2 m depth, especially if the groundwater is saline, because such water can be drawn to the surface by capillary rise.

15. Some drainage, calculated from the leaching requirement (*LR*), is essential to prevent salt buildup in irrigated soils. The *LR* increases as the electrical conductivity (*EC*), a surrogate for the salt concentration of the water, increases. Preferably the *EC* of irrigation water used for grapevines should be <0.8 dS/m.

5

The Living Soil

Soil and the Carbon Cycle

Soil is the living "skin" of the earth's terrestrial ecosystem. Like skin, it is bombarded by the sun's radiation, wind, and rain and abraded by all manner of objects scraping its surface. Unlike skin, after an initial phase of weathering, soil develops primarily from the surface downward as plant and animal residues are continually added to the surface layer. These organic residues nourish a diverse population of organisms, feeding on the dead residues and on each other. In turn, they release mineral nutrients in an interminable cycle of growth, death, and decay, commonly called the carbon (C) cycle. Figure 5.1 is a diagrammatic representation of the C cycle in a vineyard.

Green plants use energy from the sun to make carbohydrates, and subsequently proteins, lipids (fats), and other complex molecules, for their own growth and reproduction. As plant tissue matures, biochemical changes take place that lead to senescence; leaves yellow and eventually fall. In a perennial plant such as the grapevine, leaves are shed in winter but roots grow, age, and die all the time. Prunings may also be returned to the soil. Collectively, the above-ground plant material returned to the soil is called litter. Below ground, root fragments that are "sloughed off" and C compounds that leak from living roots constitute rhizo-C deposition. This below-ground C material is a readily accessible substrate (food) for microorganisms, which proliferate in the cylinder of soil surrounding each plant root, a zone called the rhizosphere. The rhizosphere is also the zone from which roots take up nutrients, as described in "The Absorbing Root," chapter 3.

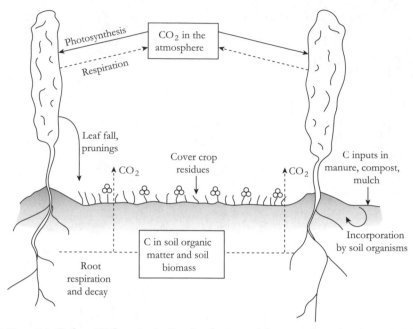

Figure 5.1 Carbon (C) flows in the C cycle of a vineyard. Inputs to the soil and vines, solid arrows; outputs as carbon dioxide (CO_2), dashed arrows.

When digesting and decomposing C substrate derived from litter and in the soil proper, organisms obtain energy and essential nutrients for growth. In so doing, in a healthy soil, they consume oxygen (O_2) and release carbon dioxide (CO_2) to the air below and above ground, thus completing the C cycle. Directly analogous to the process described for nitrogen (N) in chapter 3, C is said to be immobilized in the bodies of the soil organisms and mineralized when it is released as CO_2. The proportion of C substrate that is mineralized as CO_2 in one cycle of decomposition varies among the main groups of organisms, but the average for the whole soil is about 50%.

Individual organisms inevitably die and form the substrate for succeeding generations, which continue to decompose "old" and "new" substrates and release CO_2. The overall effect is a cascade of decomposition, with CO_2 being released and organic matter steadily being transformed into more and more recalcitrant residues (figure 5.2). The combination of living microorganisms and dead organic matter, excluding living plant roots and soil animals, is called soil organic matter (SOM).

Carbon Turnover in Soil

Steady-state equilibrium for soil C is reached when the annual input of C in organic residues above and below ground just balances the annual losses. These losses occur through decomposition, soil erosion, and some leaching of dissolved

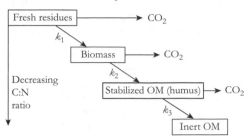

Figure 5.2 The "cascade effect" in the decomposition of organic matter (OM), starting from fresh plant residues. (White, 2003)

organic matter. Assuming the loss of C by erosion and leaching is small relative to the loss by decomposition, we may write a simple expression for the loss of C as

Loss of C (t/ha/year) $= -kC$ (5.1)

Equation 5.1 shows that C loss is the product of a rate coefficient k and the amount C of soil C present at the start of an observation period (by convention, loss is indicated by a negative sign). The coefficient k measures the fraction of soil C that decomposes during a given time period. For simplicity, if we chose a period of one year and a C input of A (measured in tonnes per hectare per year), we conclude that at steady-state equilibrium (when gains equal losses)

$A = kC$ (5.2)

and

$$\frac{C}{A} = \frac{1}{k}$$ (5.3)

Two points follow. First, the coefficient k is a weighted average of all the k values for different classes of C compound, reflecting their relative ease of decomposition, as indicated in table 5.1. Thus the average k value depends on the type of organic compounds present and their relative abundance. Figure 5.2 shows that organic residues become more and more recalcitrant as they pass through successive stages of decomposition, so the k value of well-humified organic matter is much smaller than that of fresh plant material or most microbial residues. This explains why net mineralization of N from SOM is small relative to that from legume residues, as shown in table 3.2, chapter 3, even though the C-to-N ratio of SOM is generally favorable for N mineralization.

Table 5.1 **Organic Compounds and Their Relative Ease of Decomposition under Aerobic Conditions**

Compounds and their location	Ease of decomposition	End products
Sugars and starch, present in cell contents	Very easy; large k values	CO_2, H_2O
Proteins, present in cell contents and membranes	Very easy; large k values; less easily decomposed if complexed with polyphenols	NH_4^+ and SO_4^{2-} ions, CO_2, H_2O
Organic acids including RNA and DNA, in cell contents	Easy to moderately easy; intermediate k values	CO_2, H_2O, NH_4^+, $H_2PO_4^-$ ions
Cellulose, cell wall materials	Moderately easy to difficult, depending on the organisms present; intermediate k values	Humus[a], CO_2, H_2O
Fats and oils (lipids), cell contents and membranes; waxes on leaf surfaces	Difficult to very difficult; small to very small k values	Humus, CO_2, H_2O, $H_2PO_4^-$ (from phospholipids)
Lignin and hemicellulose (cell walls); polyphenols (cell contents); chitin (insect exoskeletons, fungal cell walls)	Difficult to very difficult; small to very small k values	Humus, CO_2, H_2O, NH_4^+

Note. CO_2 = carbon dioxide; H_2O = water; NH_4^+ = ammonium; SO_4^{2-} = sulfate; $H_2PO_4^-$ = orthophosphate.

[a] Dark brown to black amorphous material formed after successive cycles of organic matter decomposition.

Second, because the dimensions of k are time^{-1}, the entity $1/k$ is a time, which defines the turnover time (in years) for C in that soil. Thus if the average k value for soil C decomposition is 0.05 units per year, the turnover time is 20 years, which is theoretically the time taken for all the C in the soil to be replaced by incoming C. In reality, some of the fresh residues may last only a few weeks, whereas other C compounds that are the end product of many cycles of decomposition may last for 1,000 years or more. Nevertheless, a soil with a C turnover time of 20 years or less is considered to be biologically active.

In a vineyard, the soil C equilibrium implicit in equation 5.2 will change if the values of A or k change. For example, rotary hoeing the mid-rows will accelerate decomposition and lead to an overall decrease in SOM. Conversely, having a cover crop of permanent grass or a mixture of herbaceous species will decrease k and increase A, leading to a gradual increase in SOM. The effect of cover crops and the addition of various kinds of organic matter to vineyard soils are discussed in "Changes in Soil Organic Matter," this chapter.

The Soil Biomass

A variety of small to microscopic organisms belonging to the plant kingdom—the "decomposers"—interact with many larger organisms, predominantly animals that live in and on the soil, called the "reducers." Collectively, this host of organisms is called the soil biomass. The biomass is concentrated in the topsoil (A horizon) because that is where the food supply is most abundant, being made up of plant and animal residues falling on the surface and C compounds released by roots.

The most important decomposers are the Archaea, bacteria, fungi, Actinobacteria (actinomycetes), and algae. There is also a group of very small animals in soil—the protozoa and nematodes. The latter do not fit neatly into either the decomposer or reducer groups because they feed predominantly on the true decomposers. Collectively, these small organisms, many of which cannot be seen with the naked eye, are referred to as the microbial biomass or, simply, as microorganisms. Measured in terms of C content, the microbial biomass ranges from 200 to 1500 kg/ha in the top 15 cm, with the larger number applying to soils under permanent grassland.

In vineyards, the most important reducers are earthworms, wood lice, mites, springtails, millipedes, centipedes, and the adult and larval stages of insects. The relationships both within and between the reducer and decomposer groups are complex, but in the context of describing how organisms influence SOM turnover we may generalize as follows.

First, although the reducers partially digest organic residues, their main benefit is in "reducing" or breaking litter and dead roots into small fragments that the decomposers can colonize and feed on. Earthworms, termites, and ants are examples of such reducers, but of these only earthworms are important in most vineyard soils. Second, life in the soil is very competitive, and, during the course of evolution, organisms have evolved that feed on other organisms. Conceptually, we recognize a hierarchical arrangement of trophic or feeding levels with plant litter, dead roots, exudates, and detritus from organisms at the bottom. Successively, going up the "food web" from this level, there are

- Archaea, bacteria, fungi, and actinomycetes that feed on dead organic matter (although small unicellular algae can be included at this level, they do not feed on dead organic matter)
- Protozoa that feed on bacteria, fungi, and algae; those mites that feed on fungi
- Omnivorous nematodes and nematodes that feed specifically on bacteria, fungi, or other nematodes
- Top predators—mites and springtails—that feed on all other organisms

At any level in this hierarchy there are also organisms that feed on (or parasitize) living plant roots. Examples of such organisms that attack grapevines are discussed in "Biological Control in Vineyard Soils," this chapter. Symbiotic, or mutually beneficial, relationships also occur. These include the fungal-root association commonly known as mycorrhizas, described in "Mycorrhizas and Nutrient Uptake," chapter 3, and the legume-*Rhizobium* symbiosis described later in box 5.2.

Overall, a viable population of decomposers and reducers is essential for many soil processes such as energy flow (from dead to living matter and back again), nutrient cycling, disease control, and the maintenance of a stable soil structure. Growers practicing organic or biodynamic viticulture believe that having a thriving soil biota is the cornerstone of a healthy soil. However, we should remember that healthy functioning of soil depends on the felicitous interaction of many physical and chemical processes with the soil biota.

Functions of the Decomposers

Decomposer organisms vary greatly in their size, numbers, and diet. Table 5.2 provides a summary of the essential features of the main groups: Archaea, bacteria, fungi, actinomycetes, and algae.

Among the decomposers, we recognize two broad nutritional groupings—the heterotrophs and autotrophs, described in table 5.3. Irrespective of these differences, however, there are some generalizations we may make about modes of nutrition and metabolism within the decomposer group.

First, microorganisms in soil are short of food most of the time, so when residues are added, the population quickly expands to consume the fresh substrate. Furthermore, if an unusual organic compound is added, microorganisms that can attack and consume that compound begin to proliferate because of their competitive advantage over other organisms that cannot.[1] When such a substrate is first added, there is a lag of several days or even weeks before it starts to disappear. However, once "enriched" with the requisite microorganisms, the soil quickly responds to further additions of that substrate, which is then rapidly decomposed. In vineyards, this behavior is important for the decomposition of some herbicides, fungicides, and pesticides that can have an adverse effect on the environment because of their chemical stability or potential to leach into runoff water. Regular and prolonged use of some fungicides and herbicides in vineyards can be deleterious to soil organisms, as discussed in "Organic, Biodynamic, and Conventional Viticulture," chapter 6.

---

[1] Because microorganisms such as bacteria have enormous "genetic plasticity," strains capable of consuming unusual substrates can evolve rapidly.

Table 5.2 Essential Features of the Main Groups Within the Decomposers

Group of microorganisms	Growth habit	Reproduction	Numbers in the soil	Physiological characteristics
Archaea	Single-cell organisms, 0.1–15 µm in size	Asexual by binary or multiple fission	Unknown	No membrane-bound nucleus; both autotrophs and heterotrophs occur (see table 5.3); first examples found in extreme environments (hot springs, highly saline, acidic or alkaline sites)
Bacteria	Single-cell organisms, 0.2–1 µm in size; favored by moist conditions	Asexual; rapid multiplication by cell division	1–4 billion/g soil; most species unidentified	Similar cell structure to Archaea but different cell biochemistry; heterotrophic and autotrophic nutrition; some parasites and symbionts; aerobes and anaerobes; collectively almost limitless variety in metabolism and ability to use diverse substrates for growth
Fungi	Some are single-celled (yeasts), but most form filaments 1–10 µm in diameter (the hyphae) that collectively form a mycelium	Asexually by cell division and sexually by producing spores on fruiting bodies	<1 million/g soil; biomass is larger than bacteria	Membrane-bound nucleus; heterotrophic; saprophytic and parasitic; some specialized symbionts as in lichens and mycorrhizas; more tolerant of soil pH(CaCl₂) <5 than most bacteria and many can decompose recalcitrant C compounds (hemicellulose, lignin, and chitin)
Actinomycetes (Actinobacteria)	Grow as networks of fine hyphae less than 0.5 µm in diameter; more delicate than the fungi	Slow growing	No reliable numbers for soil	Membrane-bound nucleus; heterotrophs; can decompose hemicellulose, lignin, and chitin; some are pathogens; some thermophilic types grow in composts at temperatures >50°C
Algae	Filamentous or single celled; grow on land and in water	Asexual and sexual	100,000–1 million/g soil	Membrane-bound nucleus; photosynthetic or heterotrophic (in the absence of light); Cyanobacteria or "blue-greens" fix atmospheric N₂ and prefer neutral to alkaline soils; green algae prefer acid soils

Compiled from Delong (1998), Angle (2000), and Thorn (2000).

Table 5.3 **Comparison of Heterotrophic and Autotrophic Microorganisms**

Heterotrophic microorganisms	Autotrophic microorganisms
Require C in organic compounds to feed on and provide C for cell growth; include the majority of species of bacteria and all fungi. Those that feed on dead organic matter are called saprophytes, whereas those feeding on living tissue are called parasites. Several parasitic bacteria and fungi are pathogens of grapevines (see "Soil Interaction in Fungal and Bacterial Diseases," this chapter).	Synthesize their cell substance from the C of CO_2 and include the remaining bacteria and most algae. Energy for this synthesis comes from sunlight (for the photosynthetic bacteria, some Archaea, and algae), or is chemical energy derived from the oxidation of inorganic compounds. Nitrifying bacteria are the most notable of the second type, one group of which gains energy from the oxidation of NH_4^+ ions and a second from the oxidation of NO_2^- (see box 3.2, chapter 3).

Note. C = carbon; CO_2 = carbon dioxide; NH_4^+ = ammonium; NO_2^- = nitrite.

Second, soil microorganisms show a general response to drastic changes such as a rapid change from air dryness to wetness or from a frozen to thawed state. When soil becomes air-dry or frozen, a substantial fraction of the microbial population dies and the remainder becomes quiescent. On rewetting or thawing, however, the bodies of the dead organisms provide an ideal substrate for the survivors. The surviving organisms and their progeny experience a flush of activity that slowly subsides as the readily decomposable material is consumed.

Third, groups of microorganisms differ in their need for O_2. Box 1.3 in chapter 1 introduced the distinction between aerobic and anaerobic conditions, but microbial metabolism is more complex in that we can recognize the following groups:

- Aerobic organisms that grow only in the presence of free O_2
- Facultative anaerobes, which normally use O_2, but can adapt to O_2-free conditions by using nitrate (NO_3^-), or other inorganic compounds such as ferric iron (Fe^{3+}), as electron acceptors in respiration, or C compounds as in fermentation
- Obligate anaerobes, which grow only in the absence of O_2 because free O_2 is toxic to them

Most fungi are aerobic, with the notable exception of yeast, which is a facultative anaerobe that ferments simple sugars predominantly to alcohol. Fungi and actinomycetes are more abundant in the litter layer because their ability to decompose lignin gives them a competitive advantage over bacteria (figure 5.3).

Fourth, soil microorganisms produce enzymes that act inside or outside their cells to digest substrates, but there are also enzymes that exist in soil independently

Figure 5.3 Mushroom fungi growing in a litter layer. (White, 2006; reprinted with permission of Wiley-Blackwell Publishing Ltd.)

of living organisms. Adsorption on soil mineral or organic matter shields these enzymes from denaturation and degradation. A common and stable soil enzyme is urease, which catalyzes the breakdown of urea in fertilizers and animal wastes.

The versatility of soil microorganisms has encouraged attempts to culture specific groups for use as biofertilizers and biostimulants. Various organic products derived from the metabolism of SOM by microorganisms are also promoted as stimulants for plant growth. Box 5.1 discusses some of these products.

Functions of the Reducers

Earthworms

Because of their size and physical activity, earthworms are more important than all the other soil invertebrates in the turnover of SOM in vineyards. As earthworms feed on dead organic matter they ingest large quantities of clay and silt-size particles and bacteria. Consequently, soil organic and mineral matter becomes more uniformly mixed and deposited in the worm feces, called casts. Sticky mucilage strengthens the casts, making them stable aggregates when dry that improve soil structure overall.

A few earthworm species live mainly in the surface soil and litter layer, provided temperature and moisture conditions are favorable. Figure 5.4 shows an example of one such species. Others burrow more deeply and deposit their casts in horizontal burrows. Still others feed on surface litter and draw it deep into the soil, to be deposited in casts in vertical burrows.

Box 5.1 Biofertilizers, Biostimulants, and Bioinoculants

In "Organic Viticulture" in chapter 3, we noted some of the organic products, commonly derived from seaweed, recommended for organic viticulture. These are usually promoted for their content of growth regulators (gibberellins, auxins, and cytokinins), sugars, and amino acids. There are also the special preparations used in biodynamic viticulture, obtained by incubating in soil a mixture of cow manure, quartz, and flower petals in cow's horn. Winegrowers who use these products claim benefits for soil and vine health, grape quality, and the distinctiveness of the wines, especially their own enjoyment of them. The products are best described as possible biostimulants for soil microorganisms, because the amounts added (as little as 90 g/ha) are far too small to have a direct effect on vine nutrition. Products derived from fishmeal are also used as biofertilizers.

Humic acid in powder form (called humate) is recommended for improving soil *CEC* and soil structure, but to have any such effect it should be added in much larger amounts than is normally recommended. Although fulvic acid is promoted as a biostimulant that can penetrate leaves, its stimulatory effect is unproven. Humic and fulvic acids do not exist as discrete organic compounds because they are a complex mixture of compounds, artificially derived from the chemical fractionation of SOM using strong alkalis and acids. The real benefit of these products in vineyards is uncertain.

Various bacteria and fungi with specific functions have been isolated from soil (e.g., *Rhizobium* bacteria; for N_2 fixation in legumes) and arbuscular mycorrhizal fungi (enhanced P uptake by roots). Other microorganisms can mineralize organic phosphate or dissolve insoluble inorganic phosphates. An example of the former is *Bacillus megatherium*, present in the commercial inoculum phosphobacterin. Although some isolates have been effective in pure culture and pot trials in glasshouses, beneficial effects are not consistently produced in the field. When introduced by seed inoculation into soil, such organisms are overwhelmed by competition from the much larger number of resident microbes. The introduced organisms do not survive, or they lose their P-solubilizing activity when they reproduce.

Of the many attempts to manipulate the rhizosphere population by the introduction of specific organisms, few have produced consistent results except for the inoculation of legume seeds with *Rhizobium* (see box 5.2). However, research in New Zealand suggests that if mulch is inoculated with the fungus *Trichoderma*, the pathogen *Botrytis* can be suppressed. In Australia, a bioinoculant called Trich-A-Soil is available that contains selected strains of this fungus, but its effectiveness is unknown.

Studies in the Barossa Valley region, South Australia, showed that the amount of SOM and the soil's chemical and physical condition influenced the species of earthworm present and their abundance. Consistent with studies on grasslands, most species are found to prefer neutral or calcareous soils, and few are found at pH less than 4.5. Under hot, dry conditions, earthworms burrow as deeply as possible into the soil and aestivate. As shown in table 5.4, extra organic matter supplied as straw mulch or from a mown cover crop, together with raising the soil

Figure 5.4 Example of shallow-burrowing earthworms collected from a moist, organic-rich soil.

pH, will markedly increase earthworm numbers. In general, earthworms are most abundant under permanent grass cover in cool, humid climates because there they are undisturbed, their food supply is plentiful, and living conditions are most favorable. Numbers may then exceed 250 worms/m^2 or a staggering 2.5 million worms/ha. Their biomass (live weight) can be as much as 3000 kg/ha.

Through earthworm activity, organic matter from the surface is mixed into the topsoil. Burrowing earthworms create pores, called biopores, as large as 5 mm in diameter that promote excellent soil drainage and aeration. Thus soil with a healthy earthworm population quickly drains to its field capacity after rain or irrigation.

Arthropods

Arthropods are a heterogeneous group of reducers that includes wood lice, mites and springtails, insects (beetles and their larvae, ants, bees, and wasps), centipedes, and millipedes. Many mites and springtails feed on plant residues and fungi in the litter, especially where thick mats build up under trees or undisturbed grassland. Predatory adult mites and springtails feed on other mites and nematodes, whereas the juveniles feed on bacteria and fungi. For example, predatory mites are effective in controlling rust mite and bud mite in several wine regions in southeastern Australia. Such biological control methods can remove the need for chemical pesticides and hence reduce the environmental side effects of viticulture.

Wood lice prefer dark, damp places and are exclusively saprophytes. Although many beetles and insect larvae that live in soil are saprophytes, others feed on living tissues and can be serious pests of grapevines. Bees, some wasps, and ants

Table 5.4 **Earthworm Biomass under Different Management Practices in a Barossa Valley Vineyard**

Management practice	Earthworm biomass (kg live weight/ha)
Mulch of grape marc	220
Ryegrass cover crop, slashed and thrown under the vines	300
Bare soil, surface application of lime	340
Straw mulch	450
Straw mulch plus lime	700

From White (2003); original data from Buckerfield and Webster (2001).

are important for the pollination of plants used as cover crops. Centipedes are fast-moving, carnivorous arthropods, feeding on insects and other small animals, whereas millipedes are slow-moving plant-feeders, mostly saprophytic. Some, however, feed on living roots, bulbs, and tubers and therefore can be pests.

Molluscs

Many molluscs (slugs and snails) feed on living plants and are therefore pests. Some species feed on fungi and the feces of other animals. Because the mollusc biomass is only 200 to 300 kg/ha in most soils, slugs and snails make only a limited contribution to the decomposition of organic matter.

Protozoa and Nematodes

Protozoa and nematodes (roundworms) are important because they control the size and composition of the bacterial and fungal populations and in so doing influence the rate of nutrient turnover. Protozoa live exclusively in water films in very small pores. Next to the protozoa, the threadlike nematodes are the smallest of the soil fauna, ranging from 0.5 to 2 mm in length. Nematodes can number up to 10 million/m^2, mainly in the litter layer and topsoil, where they feed on fungi, bacteria, protozoa, and other nematodes. Root knot and root lesion nematodes are serious parasites of grapevine roots (see "Nematodes and Their Control," this chapter).

Changes in Soil Organic Matter

How Is Soil Organic Matter Depleted?

When humans cultivate soil to grow crops, the C cycle is disturbed. Generally, in cultivated soil the C released as CO_2, plus that removed in harvested products and lost by erosion and leaching, exceeds the input from photosynthesis and the subsequent deposition of synthesized C compounds on and in the soil. A major factor contributing to this imbalance is the accelerated rate of SOM decomposition under cultivation as soil aggregates are disrupted. When aggregates break down, organic matter that was previously protected because of its inaccessibility, and even chemically recalcitrant materials, become more vulnerable to microbial attack. Nutrients such as N, phosphorus (P), and sulfur (S) that are released during SOM decomposition are taken up by plants and removed in plant products. Hence, a decrease in SOM under cultivated crops gradually leads to a decline in soil fertility. There are also undesirable changes in soil structure.

Vineyards in which the mid-rows are clean-cultivated will experience a decrease in SOM over time, especially if the soil has a light texture (sandy to sandy loam). Examples of this condition occur in the Bordeaux region, France, as illustrated in figure 5.5. Another example is shown in figure 2.16, chapter 2. On sloping land that is not terraced, continuous cultivation of the mid-rows and under vines predisposes soil to erosion, as seen in an example in the Côte d'Or region, France (figure 5.6). Although these soils are of medium texture (clay loam Brown Earths) and high in SOM in their natural state, they have been markedly depleted in C through cultivation and erosion. The next sections discuss ways of slowing or preventing SOM depletion in vineyards.

Figure 5.5 Sandy, gravely soil of low organic matter content in an overcultivated vineyard on the right bank of the Gironde River, Bordeaux region, France.

Figure 5.6 Soil erosion on the lower slopes of a clean-cultivated vineyard in the Côte d'Or, Burgundy region, France.

How to Build Up Soil Organic Matter

The opportunity to build up SOM is greater in a well-managed vineyard than under most annual crops for which the soil is cultivated and much of the crop is harvested and removed. Many winegrowers now focus on building up SOM through practices ranging from conventional viticulture with minimum soil disturbance and the use of cover crops to organic and biodynamic viticulture. Those who have adopted organic practices write with reverence about nourishing their soil, as the following example from an Australian Certified Organic vineyard in the Barossa Valley region demonstrates.

> During autumn a cover crop of legumes (beans, vetch) and cereal (oats, triticale) is sown mid-row. This is mulched off and then later rotary-hoed into the ground as a green manure crop in spring. The green manure crop provides a rich source of organic matter and nutrition for the vines. The mixed crop also ensures a diverse array of soil fungi, bacteria and earthworms are present, providing a healthy biologically active soil. "Nutri-blend," a natural fertilizer consisting of humate, basalt and soft-rock phosphate inoculated with microbes, is also used in the vineyard. Beneficial insects such as ladybirds and spiders are abundant in the vineyard.... The overall philosophy is to get it right in the vineyard with a healthy and sustainable vine and soil balance to produce sound, flavoursome grapes. (Kalleske, 2007, p. 76)

The following sections explore the advantages and disadvantages of methods for building up SOM.

Cover Crops and Mulches

The effect of a mid-row cover crop on the availability of soil water is discussed in "Cover Crops and Soil Water" in chapter 4. In addition to this effect, cover crops have important functions such as

- Protecting soil from the impact of raindrops, which initiates structural breakdown and consequent erosion
- Improving soil structure, infiltration, soil strength, and drainage
- Building up SOM and enhancing biological activity
- Providing a habitat for beneficial insects and potential predators of grapevine pests
- Suppressing weeds

The relative importance of these functions depends on the soil type and species makeup of the cover crop. A cover crop can be a single sown species, such as the winter cereal shown in figure 4.19 in chapter 4, or a mixture of species involving grasses and legumes, as seen in figure 5.7. In regions with substantial winter rainfall but dry summers, winter-growing annuals are preferred because they can be incorporated into the soil in early summer and so do not compete with the vines for water. Alternatively, perennials may be preferred in regions where water availability is not a problem but there is a need to control vine vigor or prevent soil structural decline and erosion. The deep-rooted perennial chicory can be sown in situations where improvement in soil structure is required. In organic vineyards where biodiversity is highly valued, cover crops may consist not only of a mixture of grasses and legumes but also of herbaceous broadleaf plants such as mustard (*Brassica* species), Compositae species, sages, buckwheat, and members of the carrot family.

Whatever the cover crop, mowing in early spring (at bud burst) may be necessary to reduce the risk of frost damage to the vines. An annual cover crop, especially one containing a legume, can be cultivated into the soil in spring as a "green manure" (see "Organic Viticulture," chapter 3). A legume cover crop that is incorporated when young and succulent will decompose rapidly, thus adding little to the SOM. However, its contribution to available soil N through mineralization can be appreciable, depending on how much N_2 the crop has fixed. Box 5.2 discusses the factors influencing N_2 fixation by legumes.

Cover crops potentially compete with vines for water and nutrients. In some vineyards of the Bordeaux and Dordogne regions in France, for example, a cover crop is sown in every second mid-row and alternated year by year to reduce the

Figure 5.7 A white clover–fescue cover crop late in the season in a vineyard in the Mornington Peninsula region, Victoria, Australia.

competition for water. Maintaining a cover crop during summer in low-rainfall regions is difficult where drip irrigation is used. Other possible disadvantages of a cover crop are that it may harbor pests such as the light-brown apple moth, although this is less of a problem if the cover crop is oats, faba beans, field peas, lupins, or mustard rather than other broadleaf species. There can also be a buildup of snails that attack young vines in areas sown with cover crops.

Perennial cover crops are more common now in Australian vineyards, particularly in the higher rainfall regions. Growth of the cover crop can be controlled by mowing and/or herbicides, or by sheep grazing while the vines are dormant. Figure 5.8 shows an example of the latter practice. Sometimes a cover crop may be allowed to spread under the vines, to be controlled with herbicides or cultivation when necessary. Grasses with their fibrous root systems are better for improving soil structure than legumes and generally add more organic matter. For annual grasses, regeneration is encouraged if the grasses are allowed to seed in early summer.

An under vine mulch created with mowings from a cover crop helps to suppress weeds and slowly builds up SOM. Other benefits of mulch are discussed in "Mulches, Soil Water, and Temperature" in chapter 4. Heavy applications of a mulch, such as cereal straw or composted cereal straw, can be effective in building up SOM. For example, Figure 5.9 shows the excellent result from an application of 100 m³/ha of straw composted with pig manure in a vineyard on a degraded granitic soil in the Hilltops region in New South Wales, Australia. After 18 months the A horizon, now enriched with organic matter, has developed a mellow, friable tilth. Although this heavy application of composted straw was incorporated into the soil quite quickly, uncomposted cereal straw and particularly wood chips last much longer as mulches.

Box 5.2 Factors Influencing Nitrogen Fixation by Legumes

Nitrogen (N) fixation begins when a legume root is invaded by *Rhizobium* bacteria living in the soil. This invasion stimulates cells in the root cortex to divide and enlarge to form a nodule (figure B5.2.1). Within a nodule, the bacteria synthesize an enzyme called nitrogenase that traps molecules of N_2 gas and reduces them to ammonia (NH_3), which is incorporated into amino acids.

Figure B5.2.1 Nodules on the roots of a faba bean plant. Effective nodulation usually occurs on primary roots near the base of the stem.

The reduction of NH_3 requires a large amount of energy, generated from the metabolism of carbohydrates supplied to the bacteria by the host plant. In return, the host receives amino acids that combine to form proteins. This association is called symbiotic because it benefits both the host and the *Rhizobium* bacteria. Effective N_2 fixation is indicated by a pink color inside a nodule.

Several factors determine the success of the root infection, nodule initiation, and N_2 fixation. For example,

- Low pH and low calcium (Ca) supply inhibit root infection and nodule initiation
- Competition occurs between effective and ineffective *Rhizobium* strains at the time of root infection. An ineffective strain produces many nodules that do not fix N_2, yet their presence inhibits nodulation by an effective strain

(continued)

Box 5.2 *(continued)*

- Because of the large energy demand, N_2 fixation is less likely to occur when mineral N, especially ammonium (NH_4^+), is readily available to the plant
- Inadequate carbohydrate supply and water stress decrease a nodulated root's fixation capacity
- Nodulated legumes need more phosphorus (P), molybdenum (Mo), and copper (Cu) than non-nodulated legumes and have a unique requirement for cobalt (Co)
- Nodulation and N_2 fixation can be stimulated by simultaneous infection of the roots by arbuscular mycorrhizas, mainly because of enhanced P uptake via the mycorrhiza.

When a legume requires a specific *Rhizobium* strain, it is cultured in peat and coated on the seeds with the aid of a sticky sugar solution. The seeds can also be pelleted with a coating of lime to improve *Rhizobium* survival in acid soils and the subsequent infection of legume roots.

Because of these constraints and differences in host-bacterial strain performance, the quantity of N fixed is highly variable. However, provided an effective *Rhizobium* strain is present, N fixation will be directly related to the proportion of legume in a cover crop. In vineyards where the cover crop occupies virtually all the land area, the amount of N fixed should be 50 to 100 kg N/ha/year. This fixed N is made available as nodules slough off and legume residues decompose, or through the excreta of stock animals if they graze in the vineyard. Note that this extra N may contribute to an excess vigor problem on more fertile soils.

Table 5.5 summarizes information on different types of cover crops and plant species that can be used in vineyards. Experience in California suggests that cover crops can add from 1 to 12 t dry matter/ha/year, with the larger figure applying to fully organic vineyards (McGourty, 2008).

Manures and Composts

When organic residues are consistently removed and not returned to soil, the soil becomes more acidic. Thus adding manures and compost from elsewhere can counteract soil acidification in vineyards. The nutrient value of composted manures and other composts is discussed in "Cultural Practices for Organic Viticulture" in chapter 3.

Composting involves accelerating the decomposition of plant residues or animal manures in well-aerated, moist heaps. Often plant residues and manures are co-composted to improve the nutrient content of the product, as in the example given in figure 5.9. Nonindustrial sewage sludge (biosolids) can also be composted, provided bulking-up material such as wood chips or sawdust is added. Composting is done in heaps or windrows in which the temperature should exceed 55°C for at least three days to kill weed seeds and pathogens. The heap or windrow should be turned regularly to ensure it remains aerobic and maintains an adequate temperature. Depending on the nature of the raw materials

Figure 5.8 Sheep grazing in a vineyard with a permanent grass cover crop in the Rutherglen region, Victoria, Australia.

Figure 5.9 Straw composted with pig manure has created a mellow, friable tilth in the A horizon of a degraded granitic soil.

and the ambient temperature, the process should be completed in 8 to 24 weeks. Well-made compost should be odorless and have a C-to-N ratio of less than 20. In Australia, compost should be certified according to the Australian Standard for Composts, Soil Conditioners and Mulches (AS 4454–2012). Figure 5.10 shows examples of large-scale composting operations.

Table 5.5 **Range of Species Used in Cover Crops**

Type of cover crop	Common species	Attributes	Soil factors	Limitations
Winter cereals	Oats	Suitable for green manuring or "mow and throw"; go well with legumes	Wide range of soil pH; tolerate waterlogging	Slow, early growth
	Barley	More green yield than other cereals; competitive with weeds; green manuring	Salt tolerant; prefers soil pH(CaCl$_2$) >5.5	
	Cereal rye	Does well in cool regions; drought tolerant; green manuring	Wide range of soil pH; benefits soil structure	Straw slow to decompose
	Triticale	Competes well with weeds; green manuring	Well adapted to acid soils; tolerates low fertility	Straw slow to decompose
Annual legumes	Faba bean	Winter growing; N fixing; deep tap root; green manuring; early maturing	Suited to loams and clays	Should be inoculated in acid soils; unsuitable for mowing
	Field bean	Winter growing; N fixing; green manuring	Suited to loams and clays	Shallow root system; not suitable for mowing
	Vetch	Winter–spring growth; green manuring and mowing; combines well with cereals	Wide range of soil pH	Hard seeded; may germinate when no longer needed
	Lupin	Winter growing; deep tap root; green manuring	Suited to sandy soil and low pH; intolerant of free lime	Requires a specific *Rhizobium* strain; susceptible to waterlogging
	Medics	Winter growing; drought tolerant; hard seeded; green manuring	Prefer neutral to alkaline soils	Susceptible to waterlogging

Annual grasses	Annual and Italian ryegrasses	Winter–spring growth; very competitive; fibrous root system, green manuring, and mowing; combine well with legumes	Tolerant of a range of soil conditions, except low fertility for Italian ryegrass	Annual ryegrass produces large quantities of allergenic pollen
Brassicas	Fodder radish, rape, and mustard	Winter growing; fodder radish has large tap root that breaks up hardpans; green manuring	Glucosinolates released from the roots inhibit parasitic nematodes	Large N requirement; need to be sprayed out or incorporated before seed set
Perennials	Perennial ryegrass, fescue, and cocksfoot	Suitable for cool, high-rainfall areas; mainly winter–spring growth; fibrous root system; green manuring; combine well with legumes	Prefer loams and clays; cocksfoot the most tolerant of acid soils	May compete too strongly with vines
	Strawberry clover and white clover	Winter–spring and summer (provided irrigation is supplied); N fixing; green manuring and mowing; combine well with perennial grasses	Strawberry clover tolerates adverse soil conditions better than white clover	
Australian native species	Wallaby grass and kangaroo grass	All year round growth; kangaroo grass seeds extensively in spring–early summer; possible allelopathic effect of wallaby grass on weeds	Tolerant of low fertility and acid soils	Slow to establish; high cost of seed; compete with vines for water in dry climates
	Saltbush (*Atriplex* species)	Summer growing perennials, compete with summer weeds; encourage beneficial invertebrates that control pest insects	Tolerant of salinity and low fertility	Compete with vines for water; concentrate salts in surface soil; high seed cost

Note. N = nitrogen.

From Nicholas (2004), McGourty and Regarold (2005), and Penfold (2010a, 2010b).

Compost can be incorporated into the soil at vineyard establishment, or in existing vineyards it can be spread as an under vine layer up to 5 cm thick and 50 cm wide. When applied by this method, it can have a short-term effect in moderating soil temperatures and reducing soil evaporation. However, the main benefit is that nutrients are supplied (depending on its nutrient analysis)

(a)

(b)

Figure 5.10 (A) A large compost heap under cover in a Willamette Valley vineyard, Oregon. Air is forced through the heap to improve aeration. (B) Well-made compost in windrows in the Gippsland region, Victoria, Australia.

and it can increase SOM. With respect to the latter, long-term experiments in vineyards suggest that the effects are markedly dependent on climate and soil type, and that very large additions of organic materials may be required to significantly increase SOM in some situations. For example, in the Loire Valley, France, under an annual rainfall of 525 mm, changes in SOM were measured in a clean-cultivated vineyard on sandy soil (86% sand), following additions of prunings, manure, or compost. The results in table 5.6 show that, after 28 years of large annual additions (up to 16 to 20 t/ha of fresh materials), the organic C content of the topsoil had increased to only 1.24% and 1.31%, respectively, compared with the unamended soil C content of 0.63%. As expected, gains in soil C below 30 cm were small. Clearly, in this sandy soil under low rainfall, much of added organic material decomposed each year. Nevertheless, the high turnover rate of C in the soil was beneficial for the microbial biomass, which roughly trebled in size under the manure and compost treatments.

Biochar and Similar Materials

Historically, a form of biochar was used by indigenous people in the Amazon Basin to enrich the poor soils around their settlements. The resultant dark, organic-enriched soils, called *terra preta do indio*, were rediscovered by Western soil scientists about 50 years ago and are now variously referred to as Anthropogenic or Amazonian Dark Earths. Concern about the forcing effect of greenhouse gas emissions, especially of CO_2, in warming the earth's climate has stimulated interest in recent years in the use of biochar for storing C in soil.

Table 5.6 Changes in Organic C in a Vineyard Soil (86% sand) in the Loire Valley Region, France, after 28 Years of Organic Amendments

	Topsoil (0–0.3 m)		Subsoil (0.3–0.6 m)	
Treatment	Initial organic C (%)	Final organic C (%)	Initial organic C (%)	Final organic C (%)
Control—no amendment	0.78	0.63	0.47	0.44
2 t/ha/year crushed dry prunings	0.85	0.79	0.45	0.44
20 t/ha/year fresh cow manure	0.90	1.21	0.42	0.58
16 t/ha/year spent mushroom compost	0.84	1.34	0.44	0.64

Note. C = carbon.

After Morlat and Chaussod (2008).

Biochars are made by heating organic materials ranging from wood waste to crop residues to animal manures, usually at temperatures of 350° to 500°C, in the absence of O_2. Volatile elements and organic compounds are driven off, leaving approximately 50% of the original C in a recalcitrant form, together with a variable amount of mineral ash. Biochar in agriculture is often promoted as a "win-win" material because, in addition to long-term C storage, biochars can increase cation exchange capacity (*CEC*) (thereby retaining nutrients such as Ca^{2+}, Mg^{2+}, and K^+), increase water holding capacity (in sandy soils), and improve biological function. However, according to Lehmann et al. (2011), the wide range of products with diverse physical and chemical properties, and their unknown interaction with various groups of soil organisms, raise questions about the beneficial effects of biochars. Moreover, the expectation that a biochar will remain as a C store for a very long period because of its recalcitrant nature is contrary to the expectation that it will improve soil fertility through its decomposition and release of mineral nutrients.

Experiments with biochar in vineyards are few, starting in a Swiss vineyard in 2007–2008 with trials using a charcoal-compost mixture (Niggli and Schmidt, 2010). Subsequently, trials with commercially available biochar have been established in vineyards in several European countries by the Delinat Institute (see www.ithaka-journal.net). Applied at 10t/ha, the biochar required "activation" or "charging," usually by prior mixing with a compost or manure. This reduced its C-to-N ratio, loaded the *CEC* sites with essential nutrients, and inoculated it with a range of microorganisms. Despite early positive trends, the Swiss proponents have not yet made recommendations for biochar use in viticulture, primarily because of the variation in products and their complex interaction with soil type, climate, grape variety, cover crop, and microbial colonization. In Australia, a recent biochar trial on a N-deficient soil in the King Valley, Victoria, showed no significant effect on yeast assimilable N in the harvested fruit (Winter et al., 2013), which was not surprising given that biochar was applied at only 0.7 and 1.4 t/ha.

A material similar to nonactivated biochar has been used for centuries to counteract the decline in soil C and fertility in the pale, thin soils on Chalk in the Champagne region, France. Lignite (called *cendres noires*) has been mined from local clay and silt deposits and spread in the vineyards, which are predominantly clean-cultivated.

Conclusions on How to Build Up Soil Organic Matter

Generalizing, one concludes that SOM accumulates more readily and to higher levels in cool, humid climates than in hot, dry climates. Soil type has a significant effect, with SOM accumulation being increasingly favored as clay content increases and sand content decreases. The contribution of SOM to increasing a soil's water holding capacity is significant only in sandy soils, which, paradoxically, are the soils in which increasing SOM is most difficult. Although mulches have some benefit in increasing SOM, their main purpose is to provide a protective cover for the soil, and as such they should not decompose too quickly. Regular applications of compost and manure can build up SOM, although the former is the more effective because, in the composting process, much of the easily decomposable organic matter is eliminated and more resistant material remains. Permanent grass cover crops can build up SOM and improve soil structure. However, the main effect of a cover crop with legumes is as a green manure, rather than as a contributor to SOM.

The benefits to soil health and vine performance achieved through cover crops, composts, manures, and reduced tillage derive primarily from the regular addition of decomposable organic materials and their transformation (i.e., C turnover), rather than from SOM accumulation per se.

Testing for Soil Biological Health

In the first instance, simple observations of a soil's biological health can be made in the vineyard, as outlined in box 5.3. However, in response to a growing interest in soil biology in recent years, some commercial laboratories now offer a limited array of tests, at a considerable price, for microbial activity, microbial biomass, fungi, actinomycetes, bacterial-to-fungal ratio, and identification of specific groups such as the photosynthetic and lactic acid bacteria.

From an extensive literature review, Riches et al. (2013) recommended several biological tests that should be part of a minimum data set (together with selected chemical and physical tests, reviewed by Oliver et al., 2013) to assess the overall quality of Australian vineyard soils. As shown in table 5.7, soil microbial biomass (SMB), soil organic carbon (SOC), and potentially mineralizable N (PMN) are the ones for which the most reliable benchmarking values are available, and of these SMB and PMN are highly correlated so that either could be used as an indicator. Although more sophisticated tests of biological composition and activity, such as physiological profiling (e.g., BIOLOG microplates), phospholipid fatty acid analysis, enzyme assays, and DNA-RNA sequencing, have been developed in research laboratories, very few are commercially available and none have established benchmarks for vineyard soils.

Nevertheless, as suggested for monitoring vine nutrition (see "Interpreting Plant Analyses," chapter 3), biological tests carried out from year to year can indicate how a soil's biological health is trending. Because soil biological properties are very variable in space and time (less so for SOC), when samples are collected in different years, it is important that soil moisture and temperature conditions are comparable and that the same sites are revisited. Also, because analytical methods can differ among laboratories even for the same test, the same laboratory should always be used.

Box 5.3 Field Observations for a Biologically Healthy Soil

The following simple steps can be taken to assess the soil's biological health in a vineyard:

1. Scrape away any surface litter to reveal the top of the A horizon. A litter layer that is thin and well fragmented shows that organic residues are being broken down quickly and incorporated into the soil. The topsoil should have a friable tilth and be rich in humified organic matter (a dark color). The deeper is this enriched zone, the better the soil's condition.
2. Observe the litter layer and soil surface for small insects such as wood lice and springtails. These should be plentiful and moving about actively.
3. Turn over a full spade depth of the topsoil and look for lively earthworms (see figure 5.4). More than 10 earthworm burrows per 15 × 15 × 15-cm block of soil indicates good drainage and aeration, facilitating root growth and microbial activity.

Table 5.7 **Soil Tests for Evaluating the Biological Health of Australian Vineyard Soils**

Soil test	Method of measurement	Benchmarking values for topsoil
SOC	Acid digestion (W-B) or dry combustion; W-B does not give complete digestion so the same method should be chosen for all measurements; dry combustion values need to be corrected for any free calcium carbonate; depth of soil chosen must be specified	Low <0.5 to <1.2 Medium 0.5 to 1.7 High >1 to >2 (In each class range, values increase from sandy to clay soils; values as %C[a] for 0–15 cm depth)
SMB	CFE or SIR; SIR values (in µL CO_2/g soil/hour) are multiplied by 40 to give a value comparable to SMB by CFE (Anderson and Domsch, 1978); depth of soil chosen must be specified	Low <150 Medium 150–400 High >400 (All values in mg C/kg soil for 0–15 cm depth)
PMN	Anaerobic incubation of soil for one week at 40°C; depth of soil chosen must be specified	Low <8 Optimum 8–18 High >18 (Values in mg N/kg soil/week for 0–10 cm depth)

Note. SOC = soil organic carbon; W-B = Walkley-Black; CFE = chloroform fumigation-extraction; SIR = substrate-induced respiration; SMB = soil microbial biomass; PMN = potentially mineralizable nitrogen.

Compiled from Balachandra et al. (2009), and Riches et al. (2013).

[a] To express values as soil organic matter, multiply by 1.72.

Biological Control in Vineyards

There are many soil-borne pests and disease organisms that affect grapevines directly or indirectly. Various chemicals are used to combat these pests and diseases, so it is important to know the fate of these chemicals and their residues in soil and the wider environment. Consumers are concerned about (a) the impacts of grape growing and winemaking on soil and water quality and human health, and (b) the "C footprint" of winegrowing. Consequently, winegrowers are interested in biological methods and less energy-intensive methods of controlling vineyard pests and disease, as well as the beneficial use of waste products, as discussed in "Winery Waste and the Soil" later in this chapter. Rootstocks offer one of the most effective means of specific pest control in vineyards.

The Function of Rootstocks

Varieties of *Vitis vinifera* (the scion) can be grafted onto rootstocks that are better adapted to particular soil and site conditions. Initially, the prime purpose of such grafting was to enable vinifera varieties to resist the attack of phylloxera. This pest, endemic in North America, was accidentally introduced into France in 1862, nearly devastating the wine industry and subsequently having a serious effect on wine production in many other countries. Native American *Vitis* species such as *V. rupestris* and *V. riparia* that are resistant to phylloxera were investigated as rootstocks for grafting, mainly at Montpellier in France, where a statue commemorates the rescue of the ailing French vines by vigorous American rootstocks (figure 5.11). Other native American species such as *V. champini* and *V. berlandieri* were subsequently introduced into the grafting program, and many crosses have been made using these foundation species.

The four species commonly used for rootstocks have evolved in distinct regions of North America and hence vary in many of their characteristics other than their resistance to phylloxera. Although phylloxera remains a potentially serious pest in many grape-growing regions around the world, other pests such as parasitic root nematodes are also widely recognized as a potentially serious problem. Furthermore, grafting of desirable varieties of *V. vinifera* onto a rootstock can offer other advantages such as salt tolerance, drought tolerance, tolerance of free lime ($CaCO_3$) in the soil, combating excess vigor on high-potential sites or low vigor on low-potential sites, and control of vine balance (affecting yield and fruit quality). Thus in addition to the initial four native American rootstock varieties, other varieties have been investigated in breeding and selection programs in an effort to develop commercial rootstocks with a range of desired attributes. For example, in Australia, while phylloxera and nematode resistance are important aims in current breeding programs, rootstock crosses are also being developed to

Figure 5.11 Statue at Montpellier, France, commemorating the rescue of French *Vitis vinifera* vines from phylloxera by resistant American rootstocks.

improve drought tolerance, water use efficiency, discrimination in the uptake of sodium (Na⁺), potassium (K⁺), and chloride (Cl⁻) ions, as well as vine balance and wine quality (Clingleffer et al., 2011). However, no one rootstock will satisfy all requirements, so the choice of a particular rootstock will be determined by the main factor of concern at a particular site.

In summary, growers should consider using a rootstock

- When there is a risk of phylloxera attack
- When the population of one or more parasitic nematodes in the soil is above a critical threshold level
- In salt-affected soil
- In drought-prone soil
- In a replant situation
- To manage vine growth on high or low vigor sites

Phylloxera and Its Control

Phylloxera (scientific name: *Daktulosphairia vitifoliae*) is a small aphid that feeds on vine roots and is occasionally found in leaf galls. The first above-ground sign of an attack is a premature yellowing of leaves near harvest, which is more obvious when the vine is stressed for water. Infected roots swell and become club-like, with many yellow galls (figure 5.12). Rootlet growth stops, and the uptake of water and nutrients is impaired. Progressively, there is a decline in vigor, stunted growth, and finally vine death after several years.

Because phylloxera starts from a point source and spreads outward, the organisms are most easily found on mildly affected vines at the edge of an infestation, usually in the top 30 cm of soil. Although remote sensing can detect weakened areas, these must be checked by the use of traps and soil inspection in summer (phylloxera are dormant in winter and shelter under bark on the roots). The severity and rate of spread of an infestation depend on the virulence of the phylloxera strain; there are 83 known genotypes with different degrees of lethality. In Victoria, Australia, for example, where phylloxera has existed since 1877, the largest range of genotypes occurs in the older infected areas. Less virulent genotypes may be present for 30+ years without obvious symptoms, even though the organisms are numerous on vine roots.

Phylloxera is a greater problem in poorly drained, heavy clay soils than in sandy soils. Phylloxera probably suffers from desiccation in sandy soils. For example, in the Nagambie Lakes subregion of central Victoria, a small plot of Shiraz vines survived phylloxera at the end of the 19th century because they were growing on a sandy ridge in an alluvial plain of clay loam soils (figure 5.13).

Figure 5.12 Phylloxera adults and eggs infesting a *Vitis vinifera* root. Note the swelling and yellow galls. (White, 2003. Copyright © The State of Victoria, Department of Primary Industries. Reproduced by permission of the Department of Primary Industries and therefore not an official copy. Photo by Greg Buchanan.) See color insert.

Figure 5.13 Old Shiraz vines that survived phylloxera in the 19th century on a sandy ridge in the Nagambie subregion, Victoria, Australia. (White, 2003)

Because insecticides are ineffective in controlling phylloxera, the best method of long-term control is to use cultivars that are grafted onto resistant rootstocks. Control in planting material (cuttings and rootlings) is achieved by hot-water treatment, which is also used to control nematodes, crown gall, and phytoplasma disease (see box 2.5, chapter 2). Rootstocks with *V. vinifera* parentage should not be used when strong resistance to phylloxera is required. Table 5.8 summarizes the phylloxera resistance of the more widely available rootstocks.

Nematodes and their Control

Although more than 60 species of plant-parasitic nematodes have been recorded on grapevines, many of these may have been associated with cover crops in the vineyards. Box 5.4 describes the more serious pest nematodes of grapevines. Nematodes feeding on or invading roots cause malformations and necrosis. Infested vines suffer reduced root function and loss of vigor, without any specific above-ground symptoms. Nematode damage is more severe in sandy soils. Overall, nematodes are a more serious problem than phylloxera in many Australian viticultural regions, especially in the sandier soils of the Murray Valley, which favor nematode survival and attack.

As discussed in "Preplanting" in chapter 2, fumigants are effective nematicides, especially in replant situations or where crops such as potatoes or tomatoes that harbor nematodes have been grown for several years (most nematodes infecting grapevines can attack other host plants). However, the effect of fumigation is

Table 5.8 Rootstock Crosses and Their Resistance to Phylloxera and Nematodes

Rootstock cross and main commercial varieties	Phylloxera resistance	Comments	Nematode resistance	Comments
Vitis riparia × *V. rupestris* 101–14 Millardet, Schwarzmann, 3309 and 3306 Couderc	Good	Confer low to moderate vigor on scions; more suited to cooler, wetter regions	Variable, according to nematode species	Schwarzmann has best resistance, particularly to *Meloidogyne* species, but has been found susceptible to *T. semipenetrans* and *P. vulnus* and some virulent species of *Meloidogyne* in the United States
V. berlandieri × *V. riparia* SO4, 5C Teleki, 5BB Kober, 420A Millardet	Good	Confer moderate vigor and early ripening; 5C Teleki may be too vigorous on deep sandy soils	Moderate to high, according to nematode species	SO4, 5C Teleki and 5BB Kober have better nematode resistance than 420A Millardet; poor resistance to *X. index*
V. berlandieri × *V. rupestris* 1103 Paulsen, 99 and 110 Richter, 140 Ruggeri	High	Confer moderate to high vigor (140 Ruggeri is the most vigorous); suited to drier regions	Moderate to high, according to nematode species	In Australia, 1103 Paulsen has not performed well on sandy soils with high nematode populations
V. champanii (*V. rupestris* x *V. candicans*) Ramsey, Dog Ridge, Freedom, Harmony	Good	Confer high to very high vigor, especially Ramsey and Dog Ridge on deep sandy soils that are well watered	Very high	Best adapted to hotter climates under irrigation (good wine quality requires regulated deficit irrigation); best adapted to sandy soils because of their nematode resistance; Ramsey is widely used in inland Australia
V. vinifera crosses with *V. berlandieri*, Fercal, 41 B	Poor	Rootstocks with vinifera parentage are suspect for phylloxera resistance	Poor to moderate	Not recommended, except in situations where lime-tolerance is most important
New CSIRO rootstocks, Merbein 5489, 5512 and 6262	Good	Confer low to moderate vigor; good vine balance	High	Merbein 5512 and 6262 tolerant of an aggressive strain of *M. javanica*, but Merbein 5489 is susceptible

Note. CSIRO = Commonwealth Scientific and Industrial Research Organization.

Compiled from Whiting (2003), Blieschke (2007), Walker and Stirling (2008), and Clingleffer et al. (2011).

Box 5.4 Common Nematode Pests of Grapevines

The most widely distributed nematode pest is the root knot nematode (genus *Meloidogyne*), which is found in most sandy soils in Australia, especially in the Sunraysia and Riverland regions. At least four species—*M. javanica, M. incognito, M. arenaria,* and *M. hapla*—are serious pests of grapevines in different parts of the world. The nematode invades a root, causing cells to swell and form a gall about 3 mm in diameter in which the females breed and lay eggs. A single gall with a succession of females may survive for several years. Some cover crops act as alternate hosts for these nematodes. Figure B5.4.1 shows galls on a vine root system severely infected with *M. javanica*, the most serious nematode pest in Australia.

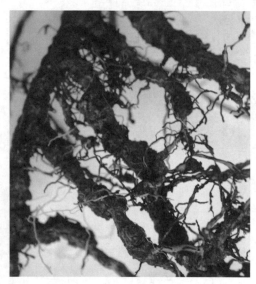

Figure B5.4.1 Grapevine roots showing galls associated with *Meloidogyne javanica* infestation. (Photo courtesy of Dr. Loothfar Rahman, National Wine and Grape Industry Centre, Wagga Wagga, New South Wales, Australia.)

The most common species of root lesion nematode (genus *Pratylenchus*) found in vineyards are *P. zeae* and *P. vulnus*; the latter is the more serious pest in the United States. Like *Meloidogyne*, these nematodes also live inside the roots. Although they burrow through the tissues without producing galls, these nematodes do cause root lesions that predispose to the invasion of fungal pathogens such as *Cylindrocarpon, Phaeomoniella,* and *Phytophthora*.

The dagger nematode (genus *Xiphinema*) is larger than other nematodes and feeds outside the roots, causing swelling of the root tips. The most common species in vineyards worldwide is *X. index*, the main vector of grape fanleaf virus. However, strains of *X. americana* are probably more widespread in Australian vineyards.

The citrus nematode (*Tylenchulus semipenetrans*) also feeds outside the roots. It is restricted to a few plant genera—most importantly, citrus and grapevines—and so is common where these two crops are grown in proximity.

not long-lasting, especially if vine planting material is not initially free of nematodes or vineyard hygiene is poor. Rootstocks provide the most effective long-term control of nematodes, although resistance may break down over time. For example, in sandy soils in the United States, resistance-breaking strains of *Meloidogyne* species have evolved from the resident soil population some 10 to 20 years after replanting (Walker and Stirling, 2008).

Table 5.8 summarizes the resistance of common rootstocks to the more problematic nematode species and strains. Because *V. champini* has a high resistance to the root knot nematode, the rootstock Ramsey became popular in the Murray-Darling region of Australia. *Muscadinia rotundifolia*, a subgenus of *Vitis*, also has strong resistance to root knot and dagger nematodes, which led to the breeding of the VR hybrids (vinifera × rotundifolia). However, their use in phylloxera-infested areas is questionable because of their vinifera parentage.

Apart from the use of rootstocks, nematode damage can be mitigated by good nursery hygiene that ensures cuttings and rootlets are nematode-free and floor management practices that encourage beneficial nematodes and depress the population of pest nematodes. In some Australian vineyards a permanent cover crop has, over time, increased the numbers of predator and omnivore nematodes and suppressed pest nematodes (Rahman et al., 2009). Similarly, species of *Brassica* (*B. napus* and *B. juncea* cv Nemfix, known as Indian mustard) produce compounds called glucosinolates that inhibit pest nematodes. These plants can be grown as cover crops and slashed before flowering, when the glucosinolate concentration is maximal, or cultivated in as green manure. Lowering the pest nematode numbers below the damage threshold, which is about 500 per kg soil for *M. javanica*, may take at least three years of such treatment (Rahman et al., 2012).

Other Rootstock Attributes

■   Salt, Drought, and pH Tolerance

Tolerance of salt and drought has become increasingly important in the choice of rootstocks, especially as competition for water to grow vines, and hence the price of water have increased in many inland regions. The threshold salinity in the root zone above which most *V. vinifera* varieties are affected is electrical conductivity (*EC*) of 1.8 dS/m, measured as the saturation extract value EC_e (see "Soil Testing for Salinity," chapter 3). As discussed in box 4.11 (chapter 4), an EC_e of 1.8 dS/m is approximately equal to an EC_{dw} of 3.6 dS/m, the latter being measurable in root-zone solutions obtained with a SoluSAMPLER (see figure 4.24). Monitoring root-zone salinity is important because wine imported into the European Union must not contain more than 394 and 606 mg/L of Na and Cl, respectively. Table 5.9 shows rootstocks that provide varying degrees of salinity tolerance, essentially through a mechanism of salt exclusion.

Table 5.9 **Tolerance of Grapevine Varieties and Rootstocks to Salinity**

Tolerance level and salinity threshold[a]	*Vitis vinifera* and common rootstocks
Sensitive, <3.6 dS/m	Most *V. vinifera* varieties on own roots, 1202 and 3309 Couderc, K51–40
Moderately sensitive, >3.6 dS/m	420A Millardet, 99 and 110 Richter, SO4, 5BB Kober,
Moderately tolerant, <6.6 dS/m	5C Teleki, 1103 Paulsen, 101–14, Ramsey
Tolerant, >6.6 dS/m	Schwarzmann, 140 Ruggeri

Compiled from Zhang et al. (2002), Whiting (2004), Roberts and Cass (2007), and Clingleffer et al. (2011).

[a] Measured as the root-zone electrical conductivity EC_{dw}.

Table 5.10 **Choice of Rootstocks for Drought Tolerance in Different Soil Types**

Soil profile characteristics[a]	Vineyard water status	Recommended rootstocks
Soil depth <20 cm; sand, loam, or clay, including any root-impeding subsoil	Dry-grown	110 Richter, 140 Ruggeri, 1103 Paulsen
	Irrigated	110 Richter, 140 Ruggeri, 1103 Paulsen, Ramsey
Soil depth 20–75 cm; sands, loams, or clays, with no root-impeding subsoil	Dry-grown	99 Richter, 140 Ruggeri, 1103 Paulsen, Ramsey, 5BB Kober
	Irrigated	99 Richter, Ramsey, 5BB Kober, 5C Teleki, Schwarzmann, SO4, 420A Millardet, 101–14 (in loams and clays)
Soil depth >75 cm; uniform or gradational profile of sand, loam, or clay	Dry-grown	99 Richter, 1103 Paulsen, Ramsey (in sand), 5BB Kober
	Irrigated	SO4, 101–14, 5C Teleki, Schwarzmann, 3306 and 3309 Couderc, 420A Millardet

Compiled from Whiting (2003), and Roberts and Cass (2007).

[a] See box 1.1, chapter 1, for soil profile descriptors.

Soil depth, texture, and the availability of irrigation water determine the choice of rootstocks for drought tolerance. Given that *V. riparia* (the river grape) naturally prefers a cool, moist environment, crosses based on this species are best suited to deep soils under cool, moist conditions. On the other hand, *V. rupestris* (the rock grape) naturally occurs in harsh habitats on rocky soils, so crosses based on this species are best suited to warm, dry climates and shallow soils. Table 5.10 groups the common rootstocks according to these characteristics.

With respect to soil pH constraints, low pH can be corrected by liming (see box 3.6, chapter 3). Although vines can grow on soils with a pH(CaCl$_2$) as low as 4, they may suffer nutritional stress, and hence reduced growth, through the effect of active aluminum (Al) and possibly manganese (Mn). On calcareous soils (pH(CaCl$_2$) 8 and above), lime-induced chlorosis (see "Micronutrient Fertilizers," chapter 3) is not easily avoided unless *V. vinifera* varieties on own roots or lime-tolerant rootstocks are used. Although rootstocks derived from *V. rupestris* and *V. riparia* are lime-sensitive, *V. berlandieri*, being a native of the limestone hills in southwest Texas and New Mexico, is the most lime-tolerant of the American species and confers this tolerance on its crosses. Of these crosses (see table 5.8), the ranking from most lime-tolerant to least is 140 Ruggeri >5BB Kober >420 Millardet >110 Richter >1103 Paulsen >99 Richter >SO4 > 5CTeleki. Rootstocks with a vinifera parentage such as 41B and Fercal are the most lime-tolerant but should not be used where phylloxera is a threat because they are very sensitive to attack by phylloxera.

■ Vine Balance and Wine Quality

The overall aims in selecting a rootstock-scion combination are as follows:

1. Avoid or minimize the influence of adverse soil factors (e.g., phylloxera, nematodes, salt, and pH)
2. Achieve a balance between root and shoot growth consistent with the yield and grape quality objectives for the vineyard

Depending on site potential, various rootstock-scion combinations range widely in performance from high to low vigor. Generally, the most vigorous vines are those with the most extensive root systems and greatest drought tolerance. Thus in table 5.10 the more vigorous rootstocks are recommended for dry-grown vineyards on shallow- to medium-depth soils of lighter texture, whereas less vigorous rootstocks are recommended for deeper soils and more favorable growing conditions. In Australian trials, the recently released Commonwealth Scientific and Industrial Research Organization rootstocks (see table 5.8) confer low to moderate vigor on Shiraz, giving higher yield-to-pruning weight ratios than popular rootstocks such as 1103 Paulsen and Ramsey.

Another factor affecting vine vigor and vine balance is the uptake of major nutrients such as N and K. Some organic-rich soils have high N availability through the natural process of mineralization, but this can be managed through the use of a grass cover crop and irrigation control (e.g., keeping the top 15–20 cm of soil dry to restrict N mineralization and the uptake of N by the vines). Rootstocks vary in their uptake of K, and this can be an important selection criterion, especially for deep sandy soils in the Murray-Darling region, Australia, that are naturally high in available K. For example, rootstocks derived from *V. champanii* (Ramsey, Freedom, and Dog Ridge) accumulate more K than 110 Richter, 1103 Paulsen, 5C Teleki, 5BB Kober, and SO4.

Soil Interaction with Fungal and Bacterial Diseases

Some grapevine disease organisms reside in soil. For others, soil management influences the severity of the disease or the actions taken for disease control affect the soil. Maintaining a species-diverse cover crop and/or mulching is helpful, if not in eliminating disease at least in controlling the incidence of disease to a degree that spraying programs can be much reduced or even made unnecessary. The benefit seems to derive from a more active and biodiverse soil microbial biomass, which is stimulated by an increase in readily available SOM and suppresses soil pathogens.

Although not soil-borne, the fungus (*Botrytis cinerea*) that causes bunch rot or gray mold survives on mummified fruits, so vineyard hygiene is important. Research in the Marlborough region, New Zealand, has shown that an under vine mulch of shredded office paper can reduce the incidence of gray mold to the extent that spraying is not needed. Mulching seems to work by improving the decomposition of vine debris, increasing competition between soil microbes and the fungus, and strengthening the skins so that the grapes are more resistant to fungal attack. Another example comes from the Griffith region, New South Wales, Australia, where the fungal root pathogen *Cylindrocarpon* spp., the cause of "black foot" in grapevines, was inhibited by green manure and compost treatments (Weckert et al., 2009).

Examples where disease control affects soil properties include the spraying of vines with wettable sulfur (S) to control powdery mildew (*Uncinula necator*). Regular use of S sprays over many years can decrease topsoil pH (see "Sulfur," chapter 3). Another example is that of downy mildew (*Plasmopara viticola*), which overwinters on leaves in the soil. Spores that are splashed onto the foliage by rain (>10 mm), when temperatures are more than 10°C for at least 24 hours, will germinate and infect the leaves. Downy mildew has traditionally been controlled by copper (Cu) sprays (Bordeaux mixture and other Cu-based fungicides), which have led to potentially toxic Cu concentrations in many topsoils in the Bordeaux region, France. Similarly, a survey of vineyard soils in Victoria, Australia, found that just over half the soils examined had topsoil Cu concentrations that exceeded Australian and New Zealand guidelines (Pietrzak, 2009).

Winery Waste and the Soil

Types of Waste and Potential Beneficial Uses

In regions as far afield as the Central Valley of California, Mendoza Province, Argentina, La Mancha, Spain, and the Murray-Darling region in Australia, winegrowers are experiencing increased competition for water from other industries and for the environment and rising water prices. Furthermore, stricter controls have been placed on the discharge of untreated winery wastewater into surface waters, to avoid polluting those waters. Consequently, the dual incentives for treatment of wastewater are

1. To recover a substantial fraction of winery wastewater treated to a level suitable for irrigation
2. To remove as much of the waste organic matter as possible, which can then be spread on pasture or composted and used in the vineyard

In addition to wastewater, another form of winery waste is marc or pomace (crushed skins, seeds, and stems), which can be composted and spread in the vine rows (see "Cultural Practices for Organic Viticulture," chapter 3). Whether the waste is in liquid or solid form, there is the potential for substantial amounts of organic matter to be returned to the soil. Box 5.5 gives examples of the amounts of organic matter that are available in liquid or solid form and their beneficial use.

Soil Factors to Consider in the Disposal of Winery Waste

The general principle to follow is to reduce, recycle, and reuse. Implementing this principle involves minimizing water use, segregating waste streams of different strength (e.g., keeping stormwater separate from wastewater), treating the water and solids recovered, reducing the amount of cleaning agents used in washing equipment, and choosing cleaning chemicals with low Na concentrations.

The variability of the waste stream is a problem for wastewater management. The biological oxygen demand (*BOD*) is determined by the amount of O_2 required by microorganisms to oxidize organic matter in the wastewater. It is expressed as a concentration of biodegradable organic matter, which ranges from 1,000 to 12,000 mg/L. Box 5.5 outlines treatment methods that are suitable for small wineries. For much larger wineries (>10,000 t crushed), settling ponds are used first to remove some of the organic matter and reduce offensive odors through anaerobic decomposition. Some of the organic N and P, along with much bacterial residue, settles in the sludge and nitrate N is lost by denitrification. These ponds are followed by aerobic ponds that are kept aerated to promote a balance between algal and bacterial growth, achieving a further reduction in *BOD*, N, and P. The effluent water from these ponds can be used to irrigate woodlots or pasture. The sludge from the ponds can be spread on agricultural fields or composted and used to build up SOM in the vineyard. Treating wastewater to a sufficiently high standard for irrigation in vineyards is expensive, costing as much as $4,000/ML in Australia.

Where wastewater is intended for land disposal, *BOD* is less of a limiting factor than the total dissolved salts (*TDS*) and sodium adsorption ratio (*SAR*) of the water (see box 2.4, chapter 2). This constraint applies equally to the use

---

Box 5.5 Amounts of Organic Waste from Wineries and Its Beneficial Use

Depending on the efficiency of the winery, a crush of 1000 t grapes produces between 1 and 3 ML wastewater. On average, the wastewater comprises 99% water and 1% solids (dissolved and particulate), most of which is organic matter. Thus for every 1000 t grapes producing, say, 2 ML wastewater, there is about 20 t of readily decomposable organic matter that can be recovered, composted, and applied to the soil. Separating the solids and water provides a source of water for irrigation.

For small wineries (<500 t crush), a favored option for water treatment is filtration through a wetland—for example, a *Phragmites* reed bed that removes much of the organic matter. The filtered water can then be used for irrigating nonvineyard land, such as pastures or woodlots, or can be blended with good-quality water at a ratio of about 1:10 and used to irrigate vines. Such filtration beds are expected to last up to 15 years before needing to be regenerated.

For intermediate-size wineries, the FILTER system developed by the Australian Commonwealth Scientific and Industrial Research Organization is a possibility (www.clw.csiro.au/publications/projects/projects20.pdf). Tile drains (perforated PVC pipe) are installed at 1-m depth and 10-m spacing in the water disposal area, which is divided into bays for flood irrigation. Much of the organic matter is retained in the soil as water percolates to the drains. However, dissolved salts also leach into the drainage water, which must be pumped into evaporation ponds. The soil under the FILTER system retains more than 80% of the wastewater N and P. The amount of water that can be applied in a season depends mainly on the soil texture. Clay soils tend to seal with the organic matter applied and may only accept up to 250 mm per year (0.25 ML/ha/year), whereas sandy soils may accept up to 1500 mm (1.5 ML/ha). In the latter case, for a 1000-t winery, about 3 ha of land would be required.

Approximately 10% of the crush goes into marc so that 100 t marc is produced per 1000 t grapes crushed. Although this material is not so readily decomposed, it can be composted at the winery site (see figure 3.15, chapter 3) and applied to the soil. In Australia, most of the marc from larger wineries is extracted to remove residual alcohol and K tartrate, dried, and pelleted for use as stock feed or soil conditioner. As a stock feed, it has almost half the nutritive value of feed-quality barley.

---

of reclaimed sewage water in vineyards. Table 5.11 gives the range of dissolved salt and nutrient concentrations in winery wastewater. Likewise, the *TDS* of reclaimed sewage water, measured by the *EC*, is usually between 1.5 and 3 dS/m.

Because sodium hydroxide (NaOH) is commonly used as a cleaning agent in a winery, the *SAR* of wastewater is raised, and, over time, this can create soil structural problems, either at the disposal site or where the water is used for productive irrigation. To counter this problem, gypsum should be applied to the soil at regular intervals. As the gypsum dissolves, Ca^{2+} ions are released and the *SAR* of

Table 5.11 **Composition of Wastewater Typical of a Large Winery**

Measured component	Value[a]
Biological oxygen demand	1000–12000 mg/L
pH	5–7.5
Electrical conductivity	1.5–3 dS/m[b]
Total nitrogen (mainly organic)	25–50 mg/L
Total phosphorus (mainly organic)	5–20 mg/L
Calcium	40–450 mg/L
Magnesium	15–60 mg/L
Sodium	40–150 mg/L
Sodium adsorption ratio	0.5–5 (mmol charge (+))$^{1/2}$

[a] The variation occurs both within and between seasons.

[b] Multiply by 640 to give total dissolved salts in milligrams per liter.

the soil solution is decreased, ideally to less than 3 (corresponding to an exchangeable sodium percentage <6). Another approach is to use K-based chemicals for cleaning and to recover much of the K as potassium tartrate from the marc (this requires the separated solids be sent away for treatment, see box 5.5).

Additionally, wastewater contains a small concentration of cationic polymers that are used in winemaking to remove suspended colloids. These polymers (possibly alum, a complex aluminum sulfate) can adversely affect soil biological health. However, if the wastewater pH is adjusted to 7 or above with lime, the polymers are precipitated; also, any zinc (Zn) and Cu contaminants are precipitated as hydroxides. Such pH adjustment can also eliminate the offensive odor of hydrogen sulfide (H_2S) gas emitted from the settling and storage ponds.

Measurements of microbial activity at long-term disposal sites (with pastures more than 20 years old) have shown that soil health is not impaired, provided the wastewater is properly treated initially. Similarly, in the McLaren Vale region, South Australia, reclaimed sewage water has been used to irrigate vineyards since 1999 and is now second only to bore water in its number of users. Tests have shown that microbial activity in these soils has not suffered compared to soils irrigated with "mains" water. Figure 5.14 shows an example of healthy vines that are drip irrigated with reclaimed water in McLaren Vale.

Leaching Requirement for Salt Control

When wastewater is used for productive irrigation, the amount applied should be determined by the *ET* demand of the crop (see "Evaporation and Transpiration," chapter 4). However, repeated applications of wastewater will gradually increase salts in the soil and restrict the range of plants that can be grown. To avoid this problem, a leaching requirement should be incorporated into the irrigation schedule, as discussed in box 4.11, chapter 4.

Figure 5.14 Healthy vines grown under irrigation with reclaimed sewage water in the McLaren Vale region, South Australia. (Photo courtesy of Dr. Belinda Rawnsley, South Australian Research and Development Institute, Adelaide, South Australia.)

Summary Points

1. The C cycle of growth, death, and decay is essential for replenishing a soil's nutrient store. Organic residues fall as litter and animal excreta on the soil surface, and dead cells and exudates are deposited around plant roots in a zone called the rhizosphere. A host of soil microorganisms colonize the organic residues, deriving energy and nutrients for growth.

2. During the decomposition process, CO_2 is produced and nutrients such as N are mineralized (as NH_4^+), depending on the C-to-N ratio of the residues. The organic residues are converted into more and more recalcitrant humic compounds. The combination of undecomposed residues, dead organisms, and humic compounds is soil organic matter (SOM).

3. Among the living soil organisms we recognize "reducers" and "decomposers." The reducers are larger organisms such as earthworms, wood lice, mites, springtails, insects and their larvae, millipedes, and centipedes that, although partially digesting organic matter, serve the important function of breaking it down into small pieces that the

decomposers can colonize. Decomposers or microorganisms are much smaller and consist of Archaea, bacteria, fungi, actinomycetes, algae, nematodes, and protozoa, which collectively have an extraordinary range of ability in decomposing organic compounds, as well as exhibiting complex predator–prey relationships.

4. Reducers and decomposers together comprise the soil biomass, amounting to as much as 2000 to 4000 kg "live weight" per hectare. The decomposers are referred to as the soil microbial biomass (SMB), which ranges in size from 100 to 1000 mg C/kg to 15-cm depth.

5. Measures of soil organic carbon (SOC) and SMB can be used to assess a soil's biological health. The benchmark for SOC (%C) ranges from >0.5 for a sand to >2 for a clay soil. The benchmark for SMB by the chloroform fumigation-extraction method is >150 to 400+ mg C/kg to 15-cm depth.

6. The food or substrate supply is the main factor controlling the size of the soil biomass. Microorganisms flourish in the rhizosphere because of the availability of C substrates from roots. Although drastic changes such as air-drying or freezing a soil kill many organisms, the survivors multiply rapidly when favorable conditions are restored because they have less competition. Microorganisms can also adapt to decomposing unusual substrates added to soil, such as organic pesticides, because in doing so they have a competitive advantage over non-adapted organisms.

7. A thriving and diverse biomass is the key to a properly functioning soil ecosystem, with benefits to plants possible from growth-regulating compounds produced by specific organisms. The mycorrhizal symbiosis between a fungus and grapevine roots improves P uptake. The legume–*Rhizobium* symbiosis "fixes" atmospheric N_2 in plants such as clovers, beans, peas, and medics that are sown in cover crops, providing N benefits to the vines.

8. Healthy soil microflora and fauna are achieved by maintaining an adequate input of organic materials through cover crops, mulches, manures, and compost. Mulches are usually organic materials that do not decompose readily, such as straw, wood chips, or biochar. A mulch moderates diurnal soil temperature fluctuations, reduces soil evaporation, and encourages earthworms.

9. Composting involves the predecomposition of organic residues, often with manure added, to reduce the C-to-N ratio. Compost is used in conventional and organic viticulture to provide nutrients and, if sufficient quantities are applied for a number of years, SOM can be increased.

10. Cover crops may consist of cereals, grass-legume mixtures, or mixtures of sown species and volunteer weeds. Cover crops create a favorable habitat for earthworms, contribute to SOM, and improve soil structure. When *Brassica* species are included they can suppress parasitic nematodes.

11. Grapevines suffer from pests such as phylloxera and nematodes that live for part or all of their life cycle in the soil. The best long-term control for these pests is to plant *V. vinifera* varieties grafted onto hybrids of resistant American rootstocks (*V. riparia, V. rupestris, V. champanii*, and *V. berlandieri*). Grafted vines are available that give greater tolerance of dry conditions, salinity, and lime and also provide better vine balance on both high- and low-potential sites.

12. Substantial volumes of liquid and solid waste are produced from a winery. Separation of the organic solids (about 1%) from water (99%) in the waste stream and treatment of the water enables it to be reused for irrigation. Solids, including marc, can be composted and applied in the vineyard. Treatment of the waste stream involves lowering its *BOD*. If the wastewater is applied to land, the *TDS* and *SAR* need to be monitored. Controlled leaching should be used to prevent a buildup of salts in the soil and gypsum applied to prevent the soil from becoming sodic.

6
Putting It All Together

Is There an Ideal Soil for Growing Wine Grapes?

In reality, there can be no generic definition of an "ideal soil" because a soil's performance is influenced by the local climate, landscape characteristics, grape variety, and cultural practices and is judged in the context of a winegrower's objectives for style of wine to be made, market potential, and profitability of the enterprise. This realization essentially acknowledges the long-established French concept of *terroir*: that the distinctiveness or typicity of wines produced in individual locations depends on a complex interaction of biophysical and human cultural factors, interpreted by many as meaning a wine's sense of place. As discussed in "Soil Variability and the Concept of *Terroir*" in chapter 1, because of this interaction of factors that determine a particular *terroir*, it is not surprising that no specific relationships between one or more soil properties and wine typicity have been unequivocally demonstrated. While acknowledging this conclusion, it is still worthwhile to examine how variations in several single or combined soil properties can influence vine performance and fruit character. These properties are:

- Soil depth
- Soil structure and water supply
- Soil strength
- Soil chemistry and nutrient supply
- Soil organisms

Soil Depth

Provided there are no subsoil constraints, the natural tendency of long-lived *Vitis vinifera*, on own roots or rootstocks, to root deeply and extensively gives it access to a potentially large store of water and nutrients. In sandy and gravely soils that are naturally low in nutrients, such as in the Médoc region of France, the Margaret River region in Western Australia, and the Wairau River plain, Marlborough region, New Zealand, the deeper the soil the better. A similar situation pertains on the deep sandy soils on granite in the Cauquenas region, Chile (figure 6.1). However, such depth may be a disadvantage where soils are naturally fertile and rain is plentiful, as in parts of the Mornington Peninsula, King and Yarra Valley regions, Victoria, Australia, and the Willamette Valley region in Oregon (see figure 1.11, chapter 1), because vine growth is too vigorous and not in balance. Although large yields can be obtained from vigorous vines, this is generally at the expense of fruit quality and the intensity of flavors in the resultant wines. When vines root deeply in such soils, their vigor should be controlled by restricting the supply of nitrogen (N) (see chapter 3), regulating irrigation (see chapter 4), growing a cover crop (permanent, if possible—see chapter 4), and through canopy management. Figure 6.2 is an example of a soil–vine combination requiring one of more of these controls.

Figure 6.1 A deep uniform soil profile formed on granite in the Cauquenas region, Chile. The pen length is 15 cm. (White, 2003)

Figure 6.2 A deep red clay loam formed on basalt in a vineyard in the Mornington Peninsula region, Victoria, Australia. The vine roots (painted white) go down to 1 m; the pit depth is 1.2 m. See color insert.

Soil depth is of less consequence in irrigated vineyards where the normal practice is to control the water content of the top 50 cm of soil only. This practice restricts the number of roots below this depth, so the effective soil depth is limited. Similarly, rooting depth is restricted in shallow soils on impermeable rock, such as is shown for vines growing on a shallow soil over limestone in figure 3.2, chapter 3. However, if the rock is fractured naturally, or by deep ripping, vine roots can grow down cracks and fissures and root more deeply. Examples occur with vines on fractured schist or shale in the Central Otago region, New Zealand, parts of Languedoc-Roussillon region, France, and the Collio del Friuli region, Italy (see figure 1.2, chapter 1). Other examples occur with vines performing well on shallow soil over limestone that is naturally fissured, as on the mid-slopes of the Côte de Nuit region in Burgundy, France (figure 6.3); the Terra Rossa of the Coonawarra region, South Australia (see figure 1.4, chapter 1); the "starfish lime stone" of the St. Emilion appellation, Bordeaux region, France; and the calcareous shale and limestone of the Monterey formation in the Paso Robles region, California. In St. Emilion and Coonawarra, the shallow limestone has in past ages been enriched with wind-blown silt-size particles, and the rate of release of stored water from soil and underlying limestone seems to ideally regulate water uptake for the production of high-quality fruit of intense flavors.

Figure 6.3 Fractured limestone rock underlying Grand Cru vineyards near Gevry-Chambertin in the Côte d'Or region, France. The pen length is 15 cm. (White, 2003)

Soil Structure and Water Supply

Soil structure interacts with soil water to determine the optimum conditions for aeration, water supply, and drainage. Fundamental to understanding this interaction is the relationship between the amount of water and the energy with which it is held, the latter being measured by the soil's matric suction at a given depth (see figure 4.13, chapter 4). Embodied in this relationship is information on the ease of water extraction by vines, the adequacy of gas exchange between the soil and air, the amount of plant available water (*PAW*), and the speed of drainage.

A soil at field capacity (*FC*) should have 10 to 15% air-filled porosity and an available water capacity (*AWC*) of 20 to 25% (see figure 4.2, chapter 4), corresponding to 200 to 250 mm water/m depth of soil. This range of air-filled porosity allows free diffusion of oxygen (O_2) into the soil to maintain aerobic respiration by roots and microorganisms and prevents too much carbon dioxide (CO_2) and other undesirable gases from accumulating in the soil. Furthermore, the soil should drain rapidly after rain or irrigation, releasing water from its large pores (macropores) that then fill with air. Such a soil does not become waterlogged (provided there is no groundwater within 2 m of the surface).

The *AWC* measures the amount of water held between *FC* and the permanent wilting point. The product of *AWC* and soil depth (in meters) determines the potential *PAW* for a given soil profile. As shown in table 4.4, chapter 4, sandy clay loams and clay loams have the most available water per centimeter of depth and consequently have the largest potential *PAW*. Achieving this potential, however, depends on the effective rooting depth of the vines. For this reason, deep-rooting vines, such as on Ramsey rootstock, are likely to be the most drought-tolerant when grown under low rainfall on deep sandy clay loams.

For irrigated vines, the depth of soil having a large proportion of readily available water (*RAW*) and deficit available water (*DAW*) can be more important than the potential *PAW*, depending on the variety and whether regulated deficit irrigation (RDI) is applied. For example, a white variety such as Chardonnay does not benefit from water stress and does best on a deep soil with large *RAW, DAW,* and *PAW*. On the other hand, a vigorous red variety such as Shiraz (Syrah) is better grown on a relatively shallow soil of smaller *RAW* and *DAW*, so that RDI can be used more effectively to manipulate berry size and grape quality (see "Managing Soil Water" later in this chapter).

In addition to these aspects of structure and water supply, soil aggregation and aggregate stability are important for optimum soil condition. Figure 4.3 in chapter 4 shows an ideal topsoil structure of predominantly crumb-like aggregates, best formed when organic matter is naturally high (>1.7% organic carbon [C] or about 3% organic matter) and well humified. However, the role of organic matter is less significant in the subsoil where the type of clay mineral and the exchangeable cations are more important. Iron (Fe) oxides (identified by their red color) and aluminum (Al) oxides contribute to stable subsoil structures, as do clay minerals that have calcium ions (Ca^{2+}) as the dominant exchangeable cation. Figure 4.4 in chapter 4 shows the ideal subsoil structure of a Terra Rossa in the Coonawarra region, South Australia, which consists of large, porous aggregates stabilized by Fe oxides, with well-defined cracks between the aggregates. The emphasis on exchangeable Ca^{2+} is necessary because too much sodium ion (Na^+) destabilizes a soil's structure. A soil's sodicity is defined by its exchangeable Na^+ content, expressed as a percentage of the cation exchange capacity, with a value of more than 6% being undesirable.

Soil Strength

Soil strength influences how easily roots can push through a soil and determines what load can be borne without structural damage, especially when wet. Soil strength, bulk density, and aggregate consistence or cohesion are interdependent. For example, bulk density (inversely related to porosity) increases as a soil becomes more compacted and, at the same time, its bulk strength, as measured by a penetrometer, increases. Soil strength for easy root penetration but good load bearing is ideally between 1 and 2 MPa when the soil is at its *FC*. The roots of *V. vinifera* can penetrate soil up to the top of this range. Variations in soil strength can also affect the availability of soil water, as expressed through the nonlimiting water range (see figure 4.8, chapter 4), an effect observed in the subsoils of a number of drip-irrigated vineyard soils in South Australia.

Although a soil may have a desirable bulk strength, it may not form a friable tilth on drying. The ideal condition is when the larger aggregates under moderate pressure break down into many smaller, stable aggregates, as seen in the upper sample in figure 6.4. Soil that is compacted because of too much wheeled traffic when wet, for example, does not break down into a friable tilth as it dries but remains massive (the lower sample in figure 6.4). A similar result can occur with soil that is compacted through natural causes, such as sodicity that induces an inherently unstable structure.

Soil Chemistry and Nutrient Supply

A healthy soil should provide a balanced supply of the essential nutrients (see table 3.1, chapter 3). Deficiencies and toxicities can be identified through soil testing. However, critical soil test values for "available" nutrients such as

Figure 6.4 Example of a friable topsoil (upper sample) and a compacted topsoil (lower sample).

phosphorus (P), potassium (K), and sulfur (S) are not specific for grapevines. They have been adapted from other orchard and agricultural crops, mainly because plant analysis has been more favored for grapevines, especially for established vineyards. Nutritional deficiencies and toxicities can be corrected with appropriate fertilizers (including organic inputs), with soil amendments (such as lime and gypsum), and in some cases by changing soil pH.

A soil's supply of N and K deserves special mention. For N, there must be an adequate supply for healthy growth, flowering, and fruit set, as well as to provide a concentration of yeast assimilable N in grape juice of 250 to 300 mg N/L for optimum fermentation. Mineral N is produced naturally through the mineralization of organic N, and table 3.2 in chapter 3 gives a qualitative estimate of the net N supply from different organic sources. When this supply is inadequate, as in sandy soils in relatively hot climates, it can be supplemented with fertilizers. However, soils of high organic C content and C-to-N ratios below 20 to 25 may produce more mineral N than a vine needs for balanced growth—the condition of excess vigor ensues. The extent to which excess vigor is a problem depends on the grape variety and the style of wine to be made. It is especially undesirable for the production of high-quality red wines, because too much shading of the grapes during ripening retards the full development of color and phenolic compounds. On the other hand, a white grape variety such as Sauvignon Blanc needs a good supply of N and water to develop the full aromatic potential of the fruit, for the making of wines meant to be consumed without ageing. Wines made from other white varieties such as Riesling and Semillon benefit from ageing and the use of fruit from less vigorous vines, provided the vines have not suffered severe water stress around veraison.

Too much N relative to K in vines, under variable weather conditions in spring, can lead to symptoms of "false K deficiency." Conversely, in soils containing illitic clay minerals, release of K^+ ions from within the clay mineral lattices can result in too much K being available and taken up by the vines. This can lead to a high juice pH, lower acidity, and color instability in red wines. Thus, especially for red grape varieties, soils with a low N mineralization potential and small proportion of illitic clay minerals are to be preferred.

Although micronutrient chemistry is complex, some generalizations can be made as to the effect of pH on micronutrient availability: that is, (a) the solubility of Fe, manganese (Mn), zinc (Zn), and copper (Cu) decreases with an increase in soil pH; (b) boron (B) availability reaches a maximum at $pH(H_2O)$ of 8, above which it decreases; and (c) molybdenum (Mo) availability is least in acid soil and increases with an increase in pH. Another consistent trend is that active Al increases as the soil becomes more acidic, resulting in an inhibition of root growth and P uptake. However, vines have survived, if not thrived, on soils of $pH(H_2O)$ as low as 4 for many years, as evidenced by the old bush vines shown in figure 6.5. By a similar reasoning that moderate water stress at a critical physiological stage

Figure 6.5 Old bush vines on a schistose soil of pH 4 in the Coteaux du Languedoc region, France. (White, 2003)

induces a distinctive wine character (discussed later), nutritional stress imposed by a very acid soil may be responsible for the so-called "flinty" taste of wines produced, for example, on the very acid soils of the Rangen region in Alsace, France.

An aspect of soil chemistry that is not directly nutritional concerns salts: a high concentration of soluble salts is undesirable. The electrical conductivity of a soil's saturation extract (EC_e) is a surrogate for the concentration of soluble salts and is easily measured. As discussed in "Salt, Drought, and pH Tolerance," chapter 5, the threshold value of EC_e is 1.8 dS/m for own-rooted vines, which is approximately equivalent to 3.6 dS/m in the soil solution in the lower root zone. If irrigation is used, the aim should be to keep the soil solution EC below the threshold (which can be higher for some rootstocks), because the salt concentration in the root zone increases over time. In soils in which vines can root deeply, groundwater should not occur within 2 m of the surface because dissolved salts can rise into the root zone by capillary action.

Soil Organisms and Organic Matter

Ideally, a soil should have a diversity of organisms, ranging from bacteria and fungi to arthropods and earthworms, which interactively decompose organic residues. Most of these organisms are saprophytes (i.e., they feed on dead organic matter) and are therefore crucial to nutrient cycling. Some, such as earthworms, burrow through the soil and improve aeration and drainage; others, such as bacteria and fungi, contribute to aggregate formation and stabilization through the secretion of gums and mucilage. An adequate content of organic matter, as indicated earlier in "Soil Structure and Water Supply," is a prerequisite for a healthy soil biota. Box 5.3 in chapter 5 describes some simple field tests to determine whether a soil is biologically healthy; table 5.7 lists laboratory tests that can be used to benchmark a soil's biological condition.

A healthy soil biota implies a population of organisms whose activities are mutually beneficial, that can suppress undesirable organisms that may be pathogenic, and that are themselves not harmful to plants. This concept is one of the underlying premises of organic (including biodynamic) viticulture— a fast-growing movement in viticulture globally. However, there is no evidence that an organic/biodynamic vineyard is any better than a well-managed conventional vineyard in suppressing parasitic nematodes and phylloxera. Hence in new and replant situations, the soil should be tested for these pests and grafted vines used as a preventative measure (see "The Function of Rootstocks," chapter 5). Furthermore, a regular input of organic residues through a cover crop, mulch, compost, or manure in conventional viticulture also encourages a diversified and metabolically active soil biota.

Managing Soil for Specific Winemaking Objectives

There is no doubt that wine grapes can be grown on a range of soils and that some soils such as deep red and brown loams and alluvial soils will provide a better growing medium for vines than others such as shallow soils on slate, schist, or hard limestone. Nevertheless, the optimum growing conditions and soil property values will inevitably be different according to the style of wine to be produced. Thus winegrowers should manage the soil in a way that enables them to achieve their specific objectives of wine style and market price point. Soil amelioration, fertilizers, water management, and biological tools can be used to modify soil behavior and the vines' response to it.

In setting objectives, the question arises: Is the wine to be an expression of place (*terroir*), or is it to conform to a brand image or "house style" that is consistent from year to year? In the former case, soil variability occurring over distances of tens of meters is a potential boon, with subtle differences in soil properties conferring typicity on the wine produced from different blocks. This approach is more likely to be followed by small winegrowers who produce distinctive wines of one variety or another from blocks of 2 to 3 ha or less. Winegrowers in the Mornington Peninsula region, Victoria, Australia, provide a good example from the New World of a conscious effort being made to relate their Pinot Noir wines to subtle differences in the soil and mesoclimate (figure 6.6). Although the distribution of soil orders in this region has been mapped at a broad scale (see figure 1.14, chapter 1), soil differences

Figure 6.6 The author explaining how soil properties in a vineyard in the Mornington Peninsula region, Victoria, Australia, could contribute to the typicity of Pinot Noir wines made from that site. (Photo courtesy of Ms. Cheryl Lee, Mornington Peninsula Vignerons Association, Victoria, Australia.)

need to be identified at a much higher resolution to demonstrate a true *terroir* effect. However, we should also recognize that, collectively, single-variety wines produced from whole regions, such as the Côte d'Or or Beaujolais in France, come to reflect, over a long period of time, a regional distinctiveness, which wine writer David Schildknecht (eRobertParker.com) calls a "strong" *terroir* effect.

The second approach to winemaking is more likely to be taken by large wine companies with extensive estates, where a blended wine is made from fruit produced on several soil types (figure 6.7). In some cases, wineries adopt both approaches, depending on the characteristics of a particular season's vintage. For example, Tim Smith (2007) of Chateau Tanunda in the Barossa Valley region, South Australia, has said "we blend different Shiraz wines to achieve a 'house style,' but wines within the blend are occasionally worthy of keeping separate to show their individuality." Single vineyard wines and wines deemed to reflect a particular *terroir* are increasingly being promoted as offering the consumer a distinctive wine experience.

Irrespective of whether grapes are grown to make a *terroir* wine or a blended varietal wine (*vin de cépage*), there are several broad issues confronting the wine industry that relate directly or indirectly to soil management. These issues relate to the "sustainability" of a winegrowing enterprise in the face of changing consumer preferences and market demands and the need to ensure that appropriate inputs are available and natural resources conserved. The main issues to be considered here are

- Managing natural soil variability in a vineyard
- Managing soil water
- Organic, biodynamic, and conventional viticulture
- Climate change and possible consequences
- Benchmarking for soil quality
- Integrated production systems (IPS) and sustainability

Managing Natural Soil Variability in a Vineyard

Natural soil variability is an important contributor to overall variability in vine performance. Precision viticulture (PV) provides a range of tools for a winegrower to manage variability and make informed management decisions, thereby gaining better control over the production system.

Chapter 1 discusses the causes of soil variability. Chapter 2 describes methods based on remote and/or proximal sensing to delineate the spatial structure of this variability. By making such measurements at a high spatial resolution and incorporating the data into a geographic information system, a winegrower can

Figure 6.7 An extensive vineyard covering more than one soil type on a gravelly floodplain in the Marlborough region, New Zealand. (Photo courtesy of Mr. Richard Merry, CSIRO Land and Water, Adelaide, South Australia.)

identify zones within a vineyard that can be managed separately, from planting through to producing fruit and ultimately harvesting. When key soil and climate variables are combined into a model, similar to the local site index described in chapter 2, which is linked to a geographic information system, the resulting map can be interpreted as defining digital *terroirs* for the site. Figure 6.8 shows the map output of such a model that has been developed for the Cowra region, New South Wales, Australia. Proffitt et al. (2006) give more examples of digital maps of variability in soil properties and vine vigor in their book *Precision Viticulture*.

Chapters 3 through 5 describe how the expression of soil variability through nutrient supply, soil structure, aeration, drainage, soil strength, water supply, salinity, sodicity, and soil biology can be recognized. Actions to remedy any onsite constraints and offsite effects of winegrowing are recommended.

Possible applications of PV include differential soil amelioration, through the use of lime or gypsum (see box 3.6, chapter 3), installing under drainage in parts prone to waterlogging (see "How to Improve Soil Drainage," chapter 4), and planning the design and layout of an irrigation system to supplement soil water (see the next section). For example, through the choice of dripper placement and delivery

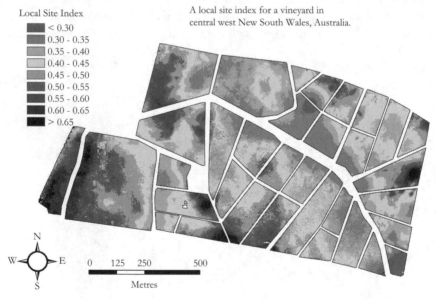

Figure 6.8 Map showing the digital *terroirs* for a vineyard in the Cowra region, New South Wales, Australia. The *terroirs* are defined by a local site index incorporating site-specific net radiation, temperature differences due to aspect and height, readily available water, clay-to-silt ratio, and root zone depth. Sites with low local site index are better suited to late varieties because of lower temperatures and frost risk. These sites also have high readily available water and are not suited to vigorous varieties like Shiraz. (Courtesy of Dr. James Taylor, Australian Centre for Precision Agriculture, Sydney, Australia.) See color insert.

rate, combined with flow control valves, water can be differentially applied to parts of a vineyard according to the available water and drainage properties of the soils. A map of vineyard soil variation showing soil textures and depths allows *RAW, DAW,* and *PAW* values to be estimated, as shown in table 4.4 in chapter 4.

None of these applications of PV will convince a winegrower of the need to act unless they result in a better product and making more profit, with minimal environmental impact. Hence, the main interest shown by winegrowers in PV is in mapping yields, vine vigor, and indices of fruit ripeness, and using this information to plan differential harvesting in time or space. To this end, the book *Precision Viticulture* (Proffitt et al., 2006) gives examples of where zones of different vine vigor within blocks were identified before harvest and the fruit in these zones was separately harvested and vinified. More recently, Bramley et al. (2011) give examples of the potential benefits to growers from the sale of their fruit to wineries, through selective harvesting based on remote sensing of vine vigor before harvest. The realization of these benefits will depend on several factors such as access to technical expertise to interpret vine vigor maps, a commitment to collecting such data for several seasons, the ability to harvest parts of blocks separately, and the willingness of wineries to pay an adequate price for higher quality fruit (Krstic, 2012).

An increasing number of small producers are adopting what might be called a *de facto* approach to PV. For example, Lloyd Brothers' vineyard of 12.5 ha of Shiraz vines in McLaren Vale region, South Australia, was planted in four blocks from 1998 to 2000 and originally treated as one commercial block with no regard to soil and other environmental variations across the blocks. Now, however, the top priority is to make small batch wines from selected rows within a single block. As Jodie Armstrong (2007) said, "The wines reflect the *terroir* of the slope with fruit from the top rows exhibiting a rich, ripe, dark berry character. The section in the middle of the slope provides more juicy fruit flavors, with rows from the bottom of the block providing a perfumed blending agent. We typically view the vineyard in three sections: super-premium, premium and table wine."

Managing Soil Water

Outside most of the Appellation d'Origine Contrôllée regions of France, vineyards have the option of managing the water supply to vines to achieve particular yield and quality objectives. As described in chapter 4, such water management can be achieved through RDI, partial root zone drying (PRD), or sustained deficit irrigation.

Research in Australia, California, and Spain has shown that a soil water deficit (*SWD*) can have a positive effect on fruit quality, especially of red grapes, but the size of the deficit (determining the degree of stress induced) and its timing must be appropriate. Expressed in terms of available water, the *SWD* should not exceed the sum of a soil's *RAW* and *DAW*. Alternatively, the soil matric suction should not exceed 100 kPa for a sandy soil or 400 kPa for a clay. In the Californian volume balance approach, the supply of water should generally not be less than 60% of full water use. Although too large a deficit may significantly reduce fruit yield in the current year, the effect is greater in the following year.

With respect to timing, the three main periods of interest are from bud burst to fruit set, fruit set to veraison, and veraison to harvest. During the first period, flowering and fruit set are the most sensitive to water stress. An *SWD* greater than the soil's *RAW* causes significant yield reductions, affecting the number of clusters per vine and the number of berries per cluster. Such stress does not normally occur in regions where the soil is at *FC* at the end of winter and is topped up by spring rains, when evaporation rates are still low to moderate. However, a possible effect of climate change may be that *SWDs* exceeding a soil's *RAW* occur more frequently early in the growing season, especially in hot inland regions, so that irrigation will be needed earlier to prevent a yield penalty. In this case, the *SWD* should be maintained within the soil's *RAW* range to restrain early vegetative growth but avoid a serious yield reduction.

Regulated deficit irrigation that imposes moderate water stress on red varieties from fruit set to veraison generally achieves the best outcome for fruit quality, although with some reduction in yield. Compared with fully watered vines, berry size decreases, soluble solids increase, malate concentration is lower, and total juice acidity is little changed. An increase in the concentration of phenolic (flavor) and color compounds is often associated with the decrease in berry size.

When RDI is prolonged to harvest, although flavor and color compounds can increase, the yield reduction is often more pronounced and the concentration of soluble solids decreased. Research at the University of California, Davis, has shown that varieties such as Cabernet Sauvignon, Merlot, and Zinfandel should be moderately stressed only until the berries reach 22° to 23° Brix, whereas for Syrah (Shiraz), the critical point is 18° to 20° Brix. Methoxypyrazine compounds that cause "green" flavors in wines reach a maximum before veraison and do not increase with subsequent watering. However, some winegrowers with well-established vines on deep loam to clay loam soils argue that "turning the taps off" after veraison is not a problem, provided that water stress does not cause shriveling of the mature berries and premature leaf fall. Implementing this approach depends on weather conditions during the post-veraison period.

Although PRD can effect similar improvements in fruit quality, the response is soil and variety dependent. Both RDI and PRD offer the added advantage of saving water, which is a critical issue for irrigated viticulture in countries as widely separated as Spain, Australia, the United States (Central Valley, California), and South Africa. Water savings up to 30% and 50% can be made with RDI and PRD, respectively. However, with RDI or PRD applied under drip irrigation, soil salinization and possibly sodicity may occur unless an adequate leaching fraction can be achieved, preferably through winter rainfall (see "Leaching, Salinity, and Sodicity Control," chapter 4).

When prolonged drought occurs and water allocations for irrigation are much reduced (as occurred during the "millennial drought" in Australia), vines may receive only 30 to 40% of the optimum irrigation volume for one or more seasons. This is an example of sustained deficit irrigation, which may maintain fruit quality but will inevitably decrease yield and can adversely affect the long-term sustainability of a vineyard.

Organic, Biodynamic, and Conventional Viticulture

Choices to be Made

Chapter 3 introduces the basic concepts and practices of organic and biodynamic (BD) viticulture. The term "conventional viticulture" covers systems in which synthetic chemicals may be used as fertilizers, as soil ameliorants, or for pest and disease control. Vineyards where both organic and conventional practices are used (ideally adopting the best of both) are sometimes called "integrated" systems. In the following discussion, organic and BD systems are considered under the generic term "organic," provided that a BD system has C inputs comparable to an organic system. This point was emphasized in "Cultural Practices for Organic Viticulture," chapter 3.

There are at least three main reasons for winegrowers to turn to organic viticulture:

1. Many wine writers and an increasing number of winegrowers claim that organic wines taste better and "capture the *terroir*" of a site.
2. Organic viticulture benefits the soil because the inputs of organic materials and absence of chemical sprays promote a healthy soil biota, which in turn enhances vine growth and fruit quality.
3. Organic viticulture is good for the environment. Losses of nutrients by leaching are minimized, and C is conserved in the soil. Greenhouse gas (GHG) emissions are indirectly reduced because chemicals and fertilizers that are made using fossil fuel energy are avoided.

Point 1 is a subjective opinion and difficult to separate from the marketing aspect of wine production. If consumers believe organic wines taste better, that is their choice. In his review of conventional, organic, and BD viticulture, Johnston (2013) concluded that insufficient scientific research has been conducted to test the claims made for the superior quality and taste of organic and BD wines and that much of what we hear is anecdotal. However, from limited studies so far in Germany, Austria, and Italy (cited by Johnston, 2013), no consistent differences have been found in juice composition or the sensory analysis of organic versus conventional or organic versus BD wines. In the United States, Ross et al. (2009) examined the sensory appreciation of wines made from organically grown vines, with or without special BD preparations, and found the differences were small and inconsistent.

Point 2 has validity in the sense that inputs of organic materials benefit the soil biota (which are generally short of food), and there are flow-on effects in improving soil structure and nutrient turnover. This point has been demonstrated in comparisons of soils under conventional and organic agriculture, in which measures of biological function (see "Testing for Soil Biological Health," chapter 5) and soil structure are generally better under organic systems, as shown in table 6.1. However, the question of whether the biological function of BD soils is superior to organic soils is unresolved because authors such as Maeder et al. (2002) in Europe concluded this was so, while others such as Reeve et al. (2005) in the United States found no significant differences in a five-year vineyard trial. The latter result is not surprising, given that BD treatments are essentially "homeopathic" because the amounts applied in any one dose are so small (e.g., 95 g/ha of preparation 500). Consequently, the chances of such a small dose of introduced microorganisms competing successfully with an overwhelming population of indigenous soil microorganisms are exceedingly small, and the indigenous organisms will rapidly decompose any growth-promoting substances in the BD preparation. With BD preparations that are sprayed onto the canopy (e.g., preparation 501 and barrel compost), especially for repeated applications, the chances of a beneficial effect on vine health are somewhat greater because small amounts of growth promotants may be absorbed through the leaves.

Aside from the effect of organic inputs, the soil biota should benefit from a reduction in, or complete absence of, chemical sprays. For example, Kremer

Table 6.1 **Mean Values of Soil Properties for Seven Biodynamic Farms Paired with Adjacent Conventional Farms Representing Mixed Enterprises on the Same Soil Types in New Zealand**

Soil property[a]	Biodynamic farms	Conventional farms
Topsoil thickness—surface and subsurface A horizons (cm)	22.8	20.6*
Bulk density (Mg/m^3)	1.07	1.15*
Penetration resistance 0–20 cm (MPa)	2.84	3.18*
Soil structure index (0–10)	7.4	5.7*
Organic C (%)	4.84	4.27*
Respiration rate (µL/g/hr)	73.7	55.4*
Potentially mineralizable N (mg/kg)	140	106*

[a] Sampling depth 0–10 cm unless otherwise indicated.
* Indicates a significant difference at the 1% probability level. Measured properties that were not significantly different are not shown.

Original data from Reganold et al. (1993).

(2009) summarized evidence for adverse effects on microbial activity and species diversity in the rhizospheres of glyphosate-resistant crops, an effect that could occur when under-vine weeds are regularly sprayed with glyphosate. Organisms such as earthworms benefit from not having potentially toxic accumulations of Cu in soil from regular applications of Bordeaux spray. New Zealand research supports the view that, with greater organic inputs from a cover crop and reduction in herbicide use, the balance between predator and pest insects and disease in the vineyard is improved. However, whether these soil changes in organic vineyards specifically result in improved fruit quality and "capture the *terroir*," compared with well-managed conventional vineyards, is debatable.

With respect to point 3, as yet there are no rigorous comparisons of the C balance between conventional and organic viticulture. Although some vineyards in France, New Zealand, Chile, and California claim to be "C neutral," this neutrality is achieved through C offsets, generally through paying for the planting of trees (buying C credits). However, more rigorous auditing (through a full life-cycle analysis) needs to be done on both the vineyard C balance and the net effect of C sequestration in trees before claims of C neutrality can be justified. In 2009, the Australian Wine Carbon Calculator (AWCC) was introduced to assist Australian wineries measure their C footprint, primarily through estimating their GHG emissions (see www.wfa.org.au). The AWCC uses internationally agreed C-accounting protocols, and its output is acceptable for reporting to Entwine Australia, the wine industry's national environmental assurance program (see later). However, the AWCC is a "work in progress," and a number of generalizations and assumptions about activities and processes must be used to complete it. For example, the impact of emissions from agricultural soils, manure application, N leaching and runoff, crop residues is not covered by the calculator, so the emission estimate is indicative only. Nevertheless, adjustment of input data allows a winegrower to do a sensitivity analysis of the business to see which changes in vineyard and winery practices have the most effect on the C footprint.

Beware—Science before Dogma!

In Australia, the National Standard for Organic and Biodynamic Produce (2009, p.50) states that an "organic or biodynamic farm must operate within a closed input system to the maximum extent possible. External farming inputs must be kept to a minimum and applied only on an 'as needs' basis." Such production is often claimed to be sustainable. However, no commercial organic vineyard is sustainable in the long term if it is completely "closed," because nutrients are exported in the grape or wine products (and prunings if they are removed from site), and some loss occurs through leaching and as gases (in the case of N). A nutrient such as P gradually reverts to less and less soluble forms in the soil, gradually becoming unavailable to the vines.

Depending on the location, there is a small input of N from the atmosphere (see "Nitrogen Cycling," chapter 3) and a variable amount from legume N_2 fixation (see box 5.2, chapter 5). Overall, however, nutrient export and loss will not be matched by inputs of nutrients from the air, nor by the vines "mining" weathering rock in the subsoil. Thus an external input of nutrients is needed for an organic vineyard to be biophysically sustainable, which means that some other part of the biosphere is depleted of nutrients (and hence set to become unsustainable). Furthermore, if an organic vineyard is truly maintained as a closed system, the inevitable depletion of nutrients over a long period will result in unthrifty vines and poor yields, making the enterprise financially nonviable and therefore unsustainable.

Although recycling C and nutrients through the return of vine residues and composted materials is to be encouraged where possible, it needs to be done in conjunction with a calculation of a vineyard's nutrient budget, as well as soil and plant testing to assess the vines' nutrient status. Furthermore, we should note that the operating costs of organic vineyards are generally higher than for conventional vineyards. Organic winegrowers in Australia at least are not paid a premium for their fruit by wineries, nor is there a consistent premium in the marketplace for organic/biodynamic wines.

Climate Change and Possible Consequences

Figure 6.9A shows that since reliable records began in 1850, according to the 11-year moving average, the mean global surface temperature has increased by about 0.8°C. The temperature trend for Australia, from 1910 to the present, is broadly similar but shows much greater interannual variability (figure 6.9B— note the differences in the scales of the x and y axes). Both graphs show an approximately linear increase in mean temperature for the last 50 years of the 20th century (up to 1999), with the increase for the planet being about 0.6°C.

The Intergovernmental Panel on Climate Change (2007) stated in its Fourth Assessment Report that the observed temperature increase (such as figure 6.9A) could be simulated only by models that incorporated anthropogenic forcing (from GHG) and not natural factors alone. The Fifth Assessment Report (IPCC, 2013) confirmed with very high confidence that this conclusion was true. Nevertheless, projection of model outputs beyond the range of measurements always introduces uncertainty not only because of an incomplete understanding of the complex interactions that influence the earth's climate but also because of assumptions made about future "emission scenarios" (Pittock, 2005), or Representative Concentration Pathways, as they are now called. As discussed in "Climate" in chapter 2, the actual temperature increase from 1970 to 2009 is tracking at the lower level of the ensemble average of model projections. Furthermore, although GHG emissions have continued to increase from 2000

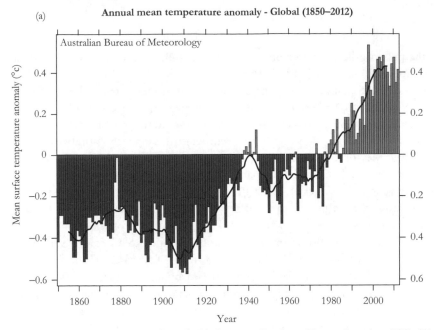

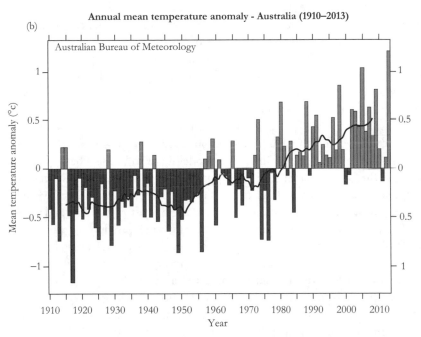

Figure 6.9 (A) Global mean annual surface temperature anomaly, 1850–2012 (baseline is the mean annual temperature for 1961–1990). The black line is the 11-year moving average. (B) Mean annual surface temperature anomaly for Australia, 1910–2013 (baseline is the mean annual temperature for 1961–1990). The black line is the 11-year moving average. (Both graphs courtesy of the Australian Bureau of Meteorology official website).

to the present, the trend in global mean surface temperature over this period has been approximately flat (see figure 6.9A).

These uncertainties in climate change trends are more marked at a regional scale, which is the scale of direct interest to winegrowers. In particular, they are concerned about the seasonality of any future temperature changes (e.g., summer versus winter), changes in the diurnal temperature range, the number of frosts and when they occur, the incidence of extreme weather events, and the amount and distribution (temporal and spatial) of rainfall. In a comprehensive review of wine geography, Jones et al. (2012) reveal the marked regional variations in temperature change during the growing season (the period of greatest change) that occurred between 1950 and 1999. For example, the observed warming in the Rhine Valley and Bordeaux regions was 0.7°C and 1.8°C, respectively, while in wine regions of northeast Spain it ranged from 1.0°C to 2.2°C. In California, Oregon, and Washington, the mean growing season temperature (GST) increased by 0.9°C, driven mostly by higher night temperatures. Given that high-quality wine production is limited to a GST in the range 13° to 21°C, some regions are now at or approaching the upper limit for their traditional varieties, as shown in figure 2.3 in chapter 2. Nevertheless, Jones et al. (2005) found that vintage quality ratings for the best wine regions increased significantly between 1950 and 1999 and that interannual variation decreased. These regional differences in past GST changes are reflected in model projections of future warming, which means that if the projections become reality, those varieties with a narrow optimum temperature range are less likely to flourish in their present regions in the future.

Global circulation or "climate" models are least reliable for projecting trends in the amounts and distribution of rainfall at regional scales. For example, for the U.S. West Coast, Lobell et al. (2006) found for a range of climate models, the projected changes in annual precipitation out to 2050 were –40 to +40% relative to the period 1960 to 1990. Simulation modeling of soil water changes can also be misleading if the spatial scale is inappropriate (White, 2013). Generalizing, Jones et al. (2012) concluded that the observed warming over the last 50 years has been largely beneficial for viticulture in many regions through longer and warmer growing seasons with less risk of frost. The unanswered question is—will the current "pause" in warming (see figure 6.9) continue, or will mean surface temperatures rise or fall over the next 10 years and more? The answer will determine whether further adaptive measures should be taken in the wine industry if the warming trend resumes, with the extension of wine-producing areas to higher latitudes and higher elevations. Other important considerations are the consequences for vine growth and balance of (a) further increases in atmospheric CO_2 concentrations and (b) increased competition for water, especially in the inland regions of California, Australia, and Spain, where irrigation is essential for viticulture to be viable. The latter consideration underlines the importance of water conservation measures such as practicing efficient irrigation, reducing soil evaporation, improving soil water holding capacity, minimizing water waste in the winery, and using reclaimed water where possible.

Benchmarking for Soil Health

Chapter 1 discusses perceptions of soil health, a term that to many in the viticulture industry, especially those promoting organic and biodynamic practices, refers primarily to a soil's biological condition. Notwithstanding the importance of biological properties, soil scientists equate soil health to soil quality. Therefore, defining soil health rests on a holistic assessment of a soil's physical, chemical, and biological properties and hence its "fitness for purpose." Nevertheless, because choice of language is important in selling a concept, if the term soil health emphasizes the need to "care" for soils, we should be prepared to adopt it.

To be confident in caring for a soil, a winegrower needs to know which soil properties are more important for soil health and what the optimum values, or range of values, are for each property against which his or her soils can be benchmarked. From a review of the physical, chemical, and biological properties most relevant to vineyard soil health, Oliver et al. (2013) and Riches et al. (2013) proposed a comprehensive minimum data set, with optimum values given where known (table 6.2). Benchmarking a vineyard requires soil to be analyzed, following standard protocols, and preferably at regular intervals to identify trends over time (see "Soil Testing," chapter 3). As more winegrowers participate and pool their data, the optimum values can be refined for broad soil classes and regions. With this knowledge, the winegrower can adjust a vineyard's management to achieve specific objectives. In many cases, the vineyard's management will then meet the soil health and environmental objectives that are the basis of sustainable winegrowing, as discussed in the next section.

Nevertheless, as discussed earlier, a winegrower's objectives can vary with the style and price point of the wines to be made. These considerations will determine the target values he or she aims for, acknowledging the French view that vines must experience a degree of soil-imposed stress to produce distinctive, quality wines (van Leeuwen, 2010).

Integrated Production Systems and Sustainability

In response to changing consumer preferences and pressure from environmental regulators, wine industries in several countries have established IPS for winegrowing. The overarching aim is to maintain the long-term sustainability of the wine industry.

Irrespective of country, the IPS have very similar principles. One of the most comprehensive systems is the Californian Sustainable Winegrowing Program (CSWP), a cooperative venture between the Wine Institute of California and the

Table 6.2 **Soil Physical, Chemical, and Biological Properties Comprising a Minimum Data Set for Soil Health Benchmarking**

Soil property and units	Suggested optimum range	Comments
Aggregate stability	<6 (semiquantitative scale from 0 [no dispersion] to 10 [high dispersion])	Aggregate stability in water; test developed by Field et al. (1997)
Air-dry consistence	2–3 (semiquantitative scale from 0 [loose] to 7 [rigid])	Method set out by the National Committee on Soil and Terrain (2009)
pH (1:5 H_2O)	5.5–8	Soil pH in 0.01M $CaCl_2$ is 0.5–0.8 units lower than in water
EC_e (dS/m)	<1.8 (higher for salt-tolerant rootstocks)	Measured in a saturated soil paste
Chloride (mg Cl/kg soil)	<10	Measured in the saturation extract
CEC (cmol (+)/kg soil)	15–50 (sands to clay soils)	Measured in an unbuffered salt solution
Exchangeable cations (percentage of CEC)— calcium, magnesium, potassium, sodium	60–80, 10–15, 1–10, <6, respectively	Measured on the same sample as CEC; if pH <5.5, measure exchangeable aluminum also
SOC (g/100 g soil)	0.5–2.0+ (lower values for sandy soils and higher values for clays)	For methods for SOC, SMB, and PMN, refer to table 5.7, chapter 5
SMB (mg C/kg soil)	150–400+	
PMN (mg N/kg soil/week)	8–18	

Note. CEC = cation exchange capacity; EC_e = electrical conductivity of the saturation extract; *SOC* = soil organic carbon; *SMB* = soil microbial biomass; *PMN* = potentially mineralizable nitrogen.
Adapted from Gugino et al. (2009), Oliver et al. (2013), and Riches et al. (2013).

California Association of Winegrape Growers (see www.wineinstitute.org). The program defines sustainable winegrowing as "winegrowing and winemaking practices that are sensitive to the environment, responsive to the needs and interests of society-at-large, and are economically feasible to implement and maintain" (Wine Institute of California, 2005–2008). The interaction of these three principles of sustainability is often illustrated by a diagram such as that in figure 6.10.

A Code of Sustainable Winegrowing Practices has been developed, which includes all aspects of winegrowing from soil and vineyard management, through pest control, energy efficiency, waste management and reduction, to wine quality, preferred purchasing, and working with neighbors. The CSWP is primarily educational, and participation is voluntary. Participants keep detailed records of their activities, which are contributed to a statewide sustainability report. In this way,

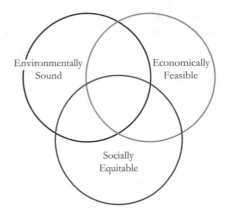

Figure 6.10 Three overlapping sustainability circles. (Graphic courtesy of the Californian Sustainable Wine Growing Alliance)

winegrowers can assess their current practice against best-management practices, enabling them to identify their strengths and weaknesses and so develop plans for improvement.

Table 6.3 shows an example of the performance categories for the organic matter criterion of soil management in the CSWP. Note the emphasis in table 6.3 is on practices that lead to an increase in soil organic matter (SOM) and hence sustainability (proceeding from category 1 to 4); but this particular criterion can be "skipped" if SOM is sufficient for a winegrower's soil type. The latter qualification illustrates the difficulty with all sustainability indices, such as SOM, in that benchmark values should be established for each environment and broad soil class, as pointed out in "Benchmarking for Soil Health" earlier in this chapter.

Examples of other countries operating a sustainability winegrowing program are New Zealand (www.nzwine.com), South Africa (www.ipw.co.za), Switzerland (www.vitiplus.ch), and Australia (www.wfa.org.au/entwineaustralia/). Certain principles are fundamental to all these systems in that self-assessment, education, training, and reporting are very important components. Although some systems have an auditing procedure and issue certificates of compliance, for others the existence and administration of environmental protection and food safety laws impose the necessary constraints on winegrowers.

The aforementioned sustainability programs vary greatly in the extent to which soil conditions and management are taken into account. For example, the program Sustainable Winegrowing New Zealand requires members to develop a soil management plan, reviewed every three to five years, that has protocols for soil sampling, soil management, and sustainability. Soil types in the vineyard are identified and soil tests carried out every three years. Much emphasis in soil management is placed on SOM and the soil's biological condition. Overall, the New Zealand program for soil and nutrient management to achieve optimum production and environmental outcomes is very comprehensive.

Table 6.3 **Performance Categories in the Organic Matter Criterion for Soil Management According to the Californian Sustainable Winegrowing Program**

Soil Management—Tilth

Criterion	Category 1	Category 2	Category 3	Category 4
Organic matter	No organic matter is added to the soil other than what the vine produces, resident vegetation is minimized in the winter, and the vineyard is clean-tilled	Resident vegetation is allowed to grow in the winter	Some form of organic matter is added to the soil annually (e.g., annual cover crop, compost, manure, or a combination of cover crop and manure or compost)	A combination of organic matter is added to the soil annually (e.g., permanent or annual cover crop, compost and/ or manure), and tillage is reduced or eliminated to lower the rate of organic matter breakdown
		Increasing sustainability		

Compiled from Jordan et al. (2004).

The Australian program, Entwine Australia, uses an industry-owned standard called the Freshcare Environmental Viticulture Code of Practice. This describes the vineyard practices required to provide customers with the assurance that produce has been grown and packed with care for the environment. The elements of Freshcare cover primarily management issues such as worker training, recording keeping, use and storage of chemicals, irrigation water, electricity and fuel consumed, and waste disposal. Land and soil are to be managed to avoid degradation, optimize SOM, and minimize on- and offsite contamination. Fertilizers and soil amendments are selected to minimize environmental impacts and their use based on plant/soil testing and crop monitoring. Although there is no reporting on soil health indicators per se, this could change in the future as better benchmarking information becomes available.

Irrespective of which sustainability program is implemented, information about the properties of soil and its management is essential to attaining the objective of sustainable wine production. Only when winegrowers take care of the soil, together with the other components of production, can this aim be achieved. Furthermore, vineyard sustainability in the biophysical sense can only be achieved when there is a balance of inputs and outputs of essential nutrients, C, and salts and avoidance of a buildup of toxic chemicals and pest organisms.

This book provides the basic knowledge of how to attain biophysical sustainability, a necessary prerequisite for realizing sustainability in the fullest sense of the term.

Summary Points

1. There is no generic definition of an "ideal soil" because a soil's performance is influenced by the local climate, landscape characteristics, grape variety, and cultural practices and is judged in the context of a winegrower's objectives for wine style, market potential, and profitability of the enterprise. For any one site, the particular combination of biophysical variables and cultural practices gives rise to its *terroir*, which may be expressed through a wine's typicity.

2. In the context of a soil's health (or quality), there are five properties or combination of properties that strongly influence a soil's performance in growing wine grapes, namely:
   - Soil depth
   - Soil structure and water supply
   - Soil strength
   - Soil chemistry and nutrient supply
   - Soil organisms

3. A deep, well-structured soil is especially important for dry-grown vines because of the need for vine roots to explore a large soil volume for nutrients and water. A topsoil structure of small (5 to 20 mm), stable aggregates is important to allow water infiltration and gas exchange with the atmosphere and to resist slaking and consequent erosion. Such a structure is best achieved with substantial inputs of organic matter or under a permanent cover crop.

4. Subsoil structure relies more on the presence of Fe and Al oxides that act as bonding agents for the soil particles. Well-defined blocky or prismatic aggregates allow good root penetration and drainage. Subsoil structure especially affects soil strength, which is a measure of a soil's load-bearing capacity and its resistance to root penetration. Measured at field capacity, the optimum range for soil strength is 1 to 2 MPa. Too high a soil strength induced by compaction, for example, decreases a soil's nonlimiting water range.

5. A healthy soil should provide a balanced supply of the macronutrients N, P, K, S, Cl, Ca, Mg, and the micronutrients Fe, Mn, Cu, Zn, Mo, and B and not show any element toxicity. Micronutrient availability is primarily controlled by soil pH, with Fe, Mn, Cu, and Zn availability decreasing with increasing pH, and Mo and B availability increasing with pH (up to 8 in the case of B). Low pH (<5 in $CaCl_2$) can predispose to Al toxicity. High concentrations of soluble salts, measured as the electrical conductivity of a saturation extract (critical value >1.8 dS/m for own-rooted vines) will adversely affect vine growth and wine quality.

6. Given that synthetic chemicals, including fertilizers, are avoided in organic and biodynamic systems, unless a nutrient budget is carried out, grapevines in these systems can suffer deficiencies in the longer term. The timing and nature of such deficiencies will depend on the soil's natural fertility, any organic inputs, and vineyard management.

7. A healthy soil biota implies a population of organisms with activities that are mutually beneficial. They can suppress undesirable organisms that may be pathogenic and are not themselves harmful to plants. These organisms should be diverse in form and function, ranging from Archaea, bacteria, actinomycetes, fungi, nematodes, and arthropods to earthworms.

8. A vineyard and its soil can be managed for specific winemaking objectives, ranging from small block, single vineyard wines that express a site's individual *terroir* to wines that are consistent in style and quality from year to year. The latter may require the blending of products from larger and more widely separated areas. Whichever objective is pursued, the winegrower may select from a variety of tools, including precision viticulture, deficit irrigation and partial root zone drying for water management, elements of organic viticulture, benchmarking soil properties against optimum values, and adopting a sustainable winegrowing program.

9. The present regions most favored for popular grape varieties may change if climate change globally and, more important, regionally continues the trend of the last 50 years of the 20th century, through higher growing season temperatures, more variable frost events, and changing patterns of rainfall. Water supplies for irrigation could become critical for hot, inland regions.

10. To promote sustainable winegrowing, several wine-producing countries have developed codes of best practice for viticulture and winemaking as part of an integrated production system. The crux of such a system is that winegrowing should be sensitive to the environment, responsive to the needs and interests of society, and economically feasible to implement and maintain—that is, to be environmentally, socially, and economically sustainable.

11. Information about the properties of soil and its best management is essential to attaining the objective of sustainable wine production. Only when winegrowers take care of the soil, together with the other components of production, can this objective be achieved.

Appendices

Appendix 1 Conversion Factors for SI (System International) Units and Non-SI Units, Including American Units, and SI Abbreviations[a] Used in This Book

To convert column 1 into column 2, multiply by	Column 1, SI unit	Column 2, non-SI unit	To convert column 2 to column 1, multiply hy
Length, area, and volume			
3.28	meter, m	foot, ft	0.304
39.4	meter, m	inch, in	0.0254
3.94×10^{-2}	millimeter, mm	inch, in	25.4
2.47	hectare, ha	acre, ac	0.405
0.265	liter, L	gallon	3.78
9.73×10^{-3}	cubic meter, m^3	acre-inch, acre-in	102.8
8.11×10^{-4}	cubic meter, m^3	acre-foot, acre-ft	1.233×10^3
35.3	cubic meter, m^3	cubic foot, ft^3	2.83×10^{-7}
0.811	megaliter, ML	acre-foot, acre-ft	1.233
Mass			
2.20×10^{-3}	gram, g	pound, lb	454
2.205	kilogram, kg	pound, lb	0.454
1.102	tonne, t	ton (U.S), ton	0.907

(continued)

243

(continued)

To convert column 1 into column 2, multiply by	Column 1, SI unit	Column 2, non-SI unit	To convert column 2 to column 1, multiply by
Quantities per unit area			
0.893	kilogram/hectare, kg/ha	pound/acre, lb/ac	1.12
0.446	tonne/hectare, t/ha	ton (U.S)/ac, ton/ac	2.24
0.107	liter/hectare, L/ha	gallon/ac	9.35
Miscellaneous			
$(9/5°C) + 32$	Celsius, °C[b]	Fahrenheit, °F	$5/9 (°F - 32)$
9.90	megapascal, MPa	atmosphere	0.101
10	Siemen/meter, S/m	millimho/centimeter, mmho/cm	0.1

[a] mega (M), $\times 10^6$; kilo (k), $\times 10^3$; deci (d), $\times 10^{-1}$; centi (c), $\times 10^{-2}$; milli (m), $\times 10^{-3}$; micro (μ), $\times 10^{-6}$; nano (n), $\times 10^{-9}$.

[b] To convert from degrees Celsius to degrees Kelvin, add 273.

Appendix 2 The Relationship between Electrical Conductivity 1 to 5 Soil-to-Water Ratio ($EC_{1:5}$) and Electrical Conductivity of the Saturation Extract (EC_e) for a Range of Soil Textures

To convert $EC_{1:5}$ values to EC_e values, multiply the $EC_{1:5}$ value by the factor K in the table A2.1. The factor decreases as texture increases because, per unit weight of soil, a clay holds more water at saturation than a sand. The dilution effect on the soil solution is therefore relatively smaller in a clay soil.

Note that these conversions are likely to overestimate EC_e in soil containing salts such as gypsum, because such salts dissolve to some extent when the soil is suspended 1 to 5 in water.

Table A2.1 **Factors to Convert $EC_{1:5}$ to EC_e Values According to Soil Texture**

Soil texture	Factor K
Loamy sand	13
Silty loam	12
Sandy loam, loam	11
Sandy clay loam, clay loam	9
Sandy clay, loamy clay, light clay	7
Medium to heavy clay	5

Adapted from White (2003).

Appendix 3 Calculating the Nitrogen Balance and Appropriate Amount of Nitrogen Fertilizer for Wine Grapes

The amount of fertilizer required is estimated from the nitrogen (N) balance equation

$$N \text{ required} = \text{Inputs} - \text{Crop removal} - \text{Losses} \tag{A3.1}$$

Table A3.1 gives some indicative results using this approach, assuming a 10 t/ha grape yield.

The first example is typical of vineyards in southeastern Australia with no mid-row cover crop, a soil carbon (C) content of 2%, and a C-to-N ratio of 12.5. Two rates of annual rainfall—600 and 1200 mm—are considered. The second example is for the same two vineyards, under the same two annual rainfalls, except that a legume cover crop is grown during the winter and cultivated in as green manure in spring. The green manure adds 1.5 t C/ha to the soil at a C-to-N ratio of 20.

As shown in figure 3.13, N fertilizer can be applied at two growth stages requiring about 2–3 g N per vine per week each time, depending on the vine density, as recommended for Western Australian vineyards. The range of 18–35 kg N/ha per season is similar to that used in the Bordeaux region, France, where an average of 30 kg N/ha per season is applied, and also for mature vines in the Willamette Valley region, Oregon. However, the N input can be two to three times greater for young vines (up to two years old) to get them established and on sandy soils of low organic matter content. In all cases, the winegrower should be prepared to adjust the fertilizer input according to plant analysis results.

Table A3.1 **Examples of N Fertilizer Required (Measured in Kilograms Nitrogen Per Hectare) Calculated from an N Balance for a Crop of 10 Tonnes Per Hectare**

Example 1: No cover crop

Atmospheric N input	Net soil mineralization	Crop requirement (shoots plus fruit) less recycling[a]	Estimated leaching plus gas losses[b]	Amount of N fertilizer required
5[c]	10	38	5	28
7.5[d]	10	38	15	35

Example 2: With a legume cover crop

Atmospheric input	Net soil mineralization, including green manure	Crop requirement (shoots plus fruit) less recycling[a]	Estimated leaching plus gas losses[b]	Amount of N fertilizer required
5[c]	25	38	10	18
7.5[d]	25	38	20	26

Note. N = nitrogen. All figures on an annual basis.

[a] Average amount of N required for growth is 76 kg N/ha (range 68–84 kg N/ha, Mullins et al., 1992), less recycling of 50% through leaf fall and prunings.

[b] Depends on soil type and rainfall (plus any irrigation).

[c,d] Annual rainfall of 600 mm and 1200 mm, respectively. Atmospheric N inputs are apportioned 50% dissolved N and 50% dry deposition, except in the higher rainfall environment where dry deposition N decreases relative to dissolved N.

References

AgNote0294. 2009. Gypsum blocks for measuring the dryness of soil.
Melbourne: Victorian Department of Environment and Primary Industries
(www.dpi.vic.gov.au/agriculture/).

Albrecht, W. A. 1975. *The Albrecht papers. Volume 1: Foundation concepts.* Kansas
City: Acres USA.

Amerine, M. A., and A. J. Winkler. 1944. Composition and quality of musts and wines of
California grapes. *Hilgardia* 15, 493–575.

Anderson, J. P. E., and K. H. Domsch. 1978. Physiological method for quantitative
measurement of microbial biomass in soil. *Soil Biology and Biochemistry* 10,
215–221.

Angle, J. S. 2000. Bacteria. In *Handbook of soil science*, ed. M. E. Sumner, C-14–C-22.
Boca Raton, FL: CRC Press.

Anon., 2013. Climate science: A sensitive matter. *The Economist* 406 (8829), 77–79.

Arbuckle, K. 2013. Increase in salinity puts growers on alert. *Australian and New Zealand
Grapegrower and Winemaker* 593, 67.

Armstrong, J. 2007. Winebiz feature of the week. *Daily Wine News*, September 1 (www.
winebiz.com.au). Adelaide: Winetitles.

Balachandra, L., R. Edis, R. E. White, and D. Chen. 2009. The relationship between
grapevine vigour and N-mineralization of soil from selected cool climate
vineyards in Victoria, Australia. *Journal of Wine Research* 20, 183–198.

Blieschke, N. 2007.Rootstock performance under drought conditions—a Yalumba
perspective. *Australian and New Zealand Grapegrower and Winemaker* 519,
38–40.

Boyer, J. D., and T. K. Wolf. 2000. Development and preliminary validation of
a Geographic Information System approach to vineyard site suitability

assessment in Virginia. *Proceedings of the 5th International Symposium on Cool Climate Viticulture and Oenology*, Workshop 16, "Site Selection and Vineyard Planning," 1–10. Adelaide: Australian Society of Viticulture and Oenology.

Bramley, R., C. Hinze, and D. Gobbett. 2010. Spatial data for improved design of vineyard (re-)planting. *Australian and New Zealand Grapegrower and Winemaker* 557, 44–53.

Bramley, R. G. V., J. Ouzman, and C. Thornton. 2011. Selective harvesting is a feasible and profitable strategy even when grape and wine production is geared towards large fermentation volumes. *Australian Journal of Grape and Wine Research* 17, 298–305.

Buckerfield, J. C., and K. A. Webster. 2001. Managing earthworms in vineyards: Improve incorporation of lime and gypsum. *Australian Grapegrower and Winemaker* 449a, 55–61.

Carbonneau, A. 2009. Evolution of canopy management: from history to scientific modelling. In *Recent advances in grapevine canopy management*, eds. N. Dokzoolian and J. Wolpert, 27–42. Davis: University of California.

Cass, A. 1999. Interpretation of some physical indicators for assessing soil physical fertility. In *Soil analysis: An interpretation manual*, eds. K. I. Peverill, L. A. Sparrow, and D. J. Reuter, 95–102. Melbourne: CSIRO Publications.

Cass, A., B. Cockcroft, and J. M. Tisdall. 1993. New approaches to vineyard and orchard soil preparation and management. In *Vineyard development and redevelopment*, ed. P. F. Hayes, 18–24. Adelaide: Australian Society of Viticulture and Oenology.

Cass, A., D. Maschmedt, and J. Chapman. 1998. Managing physical impediments to root growth. *Australian Grapegrower and Winemaker* 414, 13–17.

Chalmers, Y. M., M. Edraki, G. M. Kelly, and M. P. Krstic. 2005. The comparative response of *Vitis vinifera* L. var. Shiraz to partial root zone drying and subsurface drip irrigation. In *Proceedings of the 12th Australian Wine Industry Technical Conference*, eds. R. Blair, P. Williams, and S. Pretorius, 90–93. Urrbrae: Australian Wine Industry Technical Conference Inc.

Charlesworth, P. 2000. *Irrigation insights number 1: Soil water monitoring*. Canberra: Land and Water Australia.

Christensen, L. P. 2005. Foliar fertilization in vine mineral nutrient management programs. In *Soil environment and vine mineral nutrition*, eds. L. P. Christensen and D. R. Smart, 83–90. Davis, CA: American Society of Enology and Viticulture.

Christensen, L. P., A. N. Kasimatis, and F. L. Jensen. 1978. *Grapevine nutrition and fertilization in the San Joaquin Valley*. Oakland: Division of Agriculture and Natural Resources, University of California.

Clingleffer, P. R., B. P. Smith, H. P. Davis, E. J. Edwards, M. J. Collins, N. B. Morales, A. Boettcher, D. P. Singh, and R. R. Walker. 2011. Rootstock breeding and development for Australian conditions: Update on current work. In *Below ground management for quality and productivity: Proceedings of the Australian Society of Viticulture and Oenology and Phylloxera and Grape Industry Board of South Australia Seminar* [CD], ed. P. R. Petrie, 68–73. Adelaide: Australian Society for Viticulture and Oenology.

Davis, J. 2010. From oil exploration to winemaking. *Daily Wine News*, October 13 (www. winebiz.com.au). Adelaide: Winetitles.

Delong, E. F. 1998. Everything in moderation: Archaea as "non-extremophiles." *Current Opinion in Genetics & Development* 8, 649–654.

Dokoozlian, N. 2009. Integrated canopy management: A twenty year evolution in California. In *Recent advances in grapevine canopy management*, eds. N. Dokzoolian and J. Wolpert, 43–52. Davis: University of California.

Emerson, W. W. 1991. Structural decline in soils, assessment and prevention. *Australian Journal of Soil Research* 29, 905–921.

Fertilizer Federation Industry of Australia. (ed.). 2006. *Australian soil fertility manual.* 3rd ed. Melbourne: CSIRO Publishing.

Field, D.J., D. C. McKenzie, and A. J. Koppi. 1997. Development of an improved Vertisol stability test for SOILpak. *Australian Journal of Soil Research* 35, 843–852.

Gladstones, J. 2011. *Wine, terroir and climate change.* Kent Town: Wakefield Press.

Gladstones, J. S. 1992. *Viticulture and the environment: A study of the effects of environment on grapegrowing and wine quality with emphasis on present and future areas for growing winegrapes in Australia.* Adelaide: Winetitles.

Goldspink, B. H., and K. M. Howes. 2001. *Fertilisers for wine grapes.* 3rd ed. Bulletin 4421. Perth: Agriculture Western Australia.

Goodwin, I. 1995. *Irrigation of vineyards.* Tatura: Agriculture Victoria.

Gugino, B. K., O. J. Idowu, R. R.Schindelbeck, H. M. van Es, D. W. Wolfe, B. N. Moebius-Clune, J. E. Thies, and G. S. Abawi. 2009. *Cornell soil health assessment training manual*, edition 2.0. Geneva, NY: Cornell University.

Halliday, J. 2013. Food & wine. *The Weekend Australian Magazine*, April 13–14. Sydney: News Corporation Ltd.

Hawkins, E. 2013. Comparing global temperature observations and simulations, again (www.climate-lab-book.ac.uk/2013/), May 28, 2013.

Intergovernmental Panel on Climate Change (IPCC). 2007. Summary for policy makers. In *Climate change 2007: Synthesis report.* Fourth assessment report of the intergovernmental panel on climate change, 1–22. Geneva: IPCC.

Intergovernmental Panel on Climate Change (IPCC). 2013. Summary for policy makers. In *Climate change 2013: The physical science basis.* Fifth assessment report of the intergovernmental panel on climate change, 1–27 (www.climatechange2013. org). Geneva: IPCC.

Isbell, R. F. 1996. *The Australian soil classification.* Australian Soil and Land Survey Handbook. Melbourne: CSIRO Publishing.

Isbell, R. F. 2002. *The Australian soil classification.* Rev. ed. Australian Soil and Land Survey Handbooks Series, vol. 4. Melbourne: CSIRO Publishing.

Itami, R. M., J. Whiting, K. Hirst, and G. Maclaren. 2000. Use of analytical hierarchy process in cool climate GIS site selection for wine grapes. *Proceedings of the 5th International Symposium on Cool Climate Viticulture and Oenology*, Section 1B "Climate and Crop Estimation," 1–8. Adelaide: Australian Society of Viticulture and Oenology.

Jenny, H. 1941. *Factors of soil formation.* New York: McGraw-Hill.

Johnson, H., and J. Robinson. 2013. *The world atlas of wine.* 7th ed. London: Mitchell Beazley.

Johnston, L. 2013. *Sustainable, organic and biodynamic viticulture*. Research to Practice Manual. Adelaide: Australian Wine Research Institute Ltd.

Jones, G.V. 2006. Climate and terroir: Impacts of climate variability and change on wine. In *Fine wine and terroir: the geoscience perspective*, eds. R. W. Macqueen and L. D. Meinert, 203–216. Geoscience Canada Reprint Series No. 9. St. John's, Newfoundland: Geological Association of Canada.

Jones, G. V., R. Reid, and A. Vilks. 2012. Climate, grapes, and wine: Structure and suitability in a variable and changing climate. In *The geography of wine regions, terroir and techniques*, ed. P. H. Dougherty, 109–133. Amsterdam: Springer.

Jones, G. V., M. A. White, O. R. Cooper, and K. Storchmann. 2005. Climate change and global wine quality. *Climatic Change* 73, 319–343.

Jordan, A., J. Dlott, and K. Birdseye. 2004. The Californian wine community's code of sustainable winegrowing practices: From ground to bottle. In *Third global conference: environmental justice and global citizenship*, 1–9. Soquel, CA: SureHarvest.

Kalleske, T. 2007. Kalleske wines 2005 "Greenock" Barossa Valley Shiraz. *Australian and New Zealand Wine Industry Journal* 22 (3), 76–77.

Kay, B. D., A. P. da Silva, and J. A. Baldock. 1997. Sensitivity of soil structure to changes in organic carbon content: Predictions using pedotransfer functions. *Canadian Journal of Soil Science* 77, 655–667.

Kool, D., N. Agam, N. Lazarovitch, J. L. Heitman, T. J. Sauer, and A. Ben-Gal. 2013. A review of approaches for evapotranspiration partitioning. *Agricultural and Forest Meteorology* 184, 56–70.

Kopittke, P. M., and N. W. Menzies. 2007. A review of the use of the Basic Cation Saturation Ratio. *Soil Science Society America Journal* 71, 259–265.

Kremer, R. J. 2009. Glyphosate interactions with physiology, nutrition, and diseases of plants: Threat to agricultural sustainability? *European Journal of Agronomy* 31, 111–113.

Krstic, M. 2012. Precision management technologies prove their value in selective harvesting in Australia's major production regions. *Wine and Viticulture Journal* 27, 34–36.

Lachman, G. 2007. *Rudolph Steiner: An introduction to his life and work*. New York: J. P. Tarcher/Penguin.

Lehman, J., M. C. Rillig, J. Thies, C. A. Masiello, W. C. Hockaday, and D. Crowley. 2011. Biochar effects on soil biota—a review. *Soil Biology and Biochemistry* 43, 1812–1836.

Letey, J. 1985. Relationship between soil physical properties and crop production. In *Advances in soil science*, Vol. 1, ed. B.A. Stewart, 277–294. New York: Springer Verlag.

Lobell, D. B., C. B. Field, K. N. Cahill, and C. Bonfils. 2006. Impacts of future climate change on California perennial crop yields: Model projections with climate and crop uncertainties. *Agricultural and Forest Meteorology* 141, 208–218.

López-Urrea, R., A. Montoro, F., Manas, P., López-Fuster, and E. Fereres. 2012. Evapotranspiration and crop coefficients from lysimeter measurements of mature "Tempranillo" wine grapes. *Agricultural Water Management* 112, 13–20.

Maeder, P., A. Fliebbach, D. Dubois, L. Gunst, P. Fried, and U. Niggli 2002. Soil fertility and biodiversity in organic farming. *Science* 296, 1694–1697.

Maltman, A. 2009. On vineyard geology and "minerality" of wine. In *The relationship of geology, soils, hydrology, and climate to wine: A special tribute to George Moore.* Abstracts with programs 41 (7) 695. Portland, OR: Geological Society of America.

Maschmedt, D. J., R. W. Fitzpatrick, and A. Cass. 2002. *Key for identifying categories of vineyard soils in Australia.* Technical Report No. 30/02. Adelaide: CSIRO Land and Water.

McGourty, G. T. 2008. Growers transition to organic: The ins and outs of registration, certification and building your soil. *Wines and Vines* 89 (7), 26–36.

McGourty, G. T., and J. P. Reganold. 2005. Managing vineyard organic matter with cover crops. In *Soil environment and mineral nutrition,* eds. L. P. Christensen and D. R. Smart, 145–151. Davis, CA: American Society of Enology and Viticulture.

McKenzie, D. C. 2000. Soil survey options prior to vineyard design. *Australian Grapegrower and Winemaker* 438a, 144–151.

Morlat, R., and R. Chaussod. 2008. Long-term additions of organic amendments in a Loire Valley vineyard. I: Effects on properties of a calcareous sandy soil. *American Journal of Enology and Viticulture* 59, 353–363.

Mullins, M, G., A. Bouquet, and L. E. Williams. 1992. *Biology of the grapevine.* Cambridge: Cambridge University Press.

Murray, R. 2010. Vine roots growing in a poor environment. *Australian and New Zealand Grapegrower and Winemaker* 562, 46–47.

National Committee on Soil and Terrain. 2009. *Australian soil and land survey field handbook.* 3rd ed., 186–187, Collingwood, Victoria: CSIRO Publishing.

National Standard for Organic and Biodynamic Produce. 2009. 3.4 ed. (www.daff. gov.au/biosecurity/export/organic-biodynamic). Canberra: Department of Agriculture.

Nicholas, P. (ed.). 2004. *Soil, irrigation and nutrition.* Grape Production Series No. 2. Adelaide: South Australian Research and Development Institute.

Niggli, C., and H.-P. Schmidt. 2010. Biochar in vineyards. *Ithaka Journal* 1, 318–322.

Oliver, D. P., R. G. V. Bramley, D. Riches, I. Porter, and J. Edwards. 2013. Review: Soil physical and chemical properties as indicators of soil quality in Australian viticulture. *Australian Journal of Grape and Wine Research* 19, 129–139.

Otto, A., F. E. L. Otto, O. Boucher, J. Church, G. Hegerl, P. M. Forster, N. P. Gillett, J. Gregory, G. C. Johnson, R. Knutti, N. Lewis, U. Lohmann, J. Marotzke, G. Myhre, D. Shindell, B. Stevens, and M. R. Allen. 2013. Energy budget constraints on climate response. *Nature Geoscience* 6 (6), 415–416.

Penfold, C. 2010a. Native grass cover crops. *Australian and New Zealand Grapegrower and Winemaker* 554, 48–50.

Penfold, C. 2010b. Should I try saltbush for a cover crop? *Australian and New Zealand Grapegrower and Winemaker* 552, 16–18.

Pietrzak, U. 2009. The fate and behaviour of copper in vineyards soils of Victoria, Australia. PhD thesis, Monash University.

Pittock, A. B. 2005. *Climate change: turning up the heat.* Collingwood, Victoria: CSIRO Publishing.

Prichard, T., B. Hanson, L. Schwank, P. Verdegaal, and R. Smith. 2004. *Deficit irrigation of quality winegrapes using micro-irrigation techniques.* Davis: Department of Land, Air and Water Resources, University of California.

Prichard, T., and P. S. Verdegaal. 2001. *Effect of water deficit on winegrape yield and quality.* Davis: Department of Land, Air and Water Resources, University of California.

Proffitt, T., E. Jarvis, and M. Gibberd. 2013. A review on the use of cover crops to control vigorous vine growth. *Australian and New Zealand Grapegrower and Winemaker* 594, 29–34.

Proffitt, T., R. Bramley, D. Lamb, and E. Winter. 2006. *Precision viticulture: A new era in vineyard management and wine production.* Adelaide: Winetitles.

Rahman, L., M. A. Whitelaw-Weckert, and G. Dunn. 2012. Floor management practices to reduce pest-nematodes. *Australian and New Zealand Grapegrower and Winemaker* 577, 20–23.

Rahman, L., M. A. Whitelaw-Weckert, R. J. Hutton, and B. Orchard. 2009. Impact pf floor vegetation on the abundance of nematode trophic groups in vineyards. *Applied Soil Ecology* 42, 96–106.

Reeve, J. R., L. Carpenter-Boggs, J. P. Reganold, A. L. York, G. McGourty, and L. P. McCloskey. 2005. Soil and winegrape quality in biodynamically and organically managed vineyards. *American Journal of Enology and Viticulture* 56, 367–376.

Reganold, J. P., A. S. Palmer, J. C. Lockhart, and A. N. Macgregor. 1993. Soil quality and financial performance of biodynamic and conventional farms in New Zealand. *Science* 260, 344–349.

Riches, D., I. J. Porter, D. P. Oliver, R. G. V. Bramley, B. Rawnsley, J. Edwards, and R. E. White. 2013. Review: Soil biological properties as indicators of soil quality in Australian viticulture. *Australian Journal of Grape and Wine Research* 19, 311–323.

Roberts, D., and A. Cass 2007. Influence of available water and salinity on rootstock selection. *Practical Winery and Vineyard* 29 (2), 34–42.

Robinson, J. B. 2005. Critical plant tissue values and application of nutritional standards for practical use in vineyards. In *Soil environment and vine mineral nutrition,* eds. L. P. Christensen and D. R. Smart, 61–68. Davis, CA: American Society of Enology and Viticulture.

Robinson, J. B., M. T. Treeby, and R. A. Stephenson. 1997. Fruits, vines and nuts. In *Plant analysis: An interpretation manual.* 2nd ed., eds. D. J. Reuter and J. B. Robinson, 349–382. Melbourne: CSIRO Publishing.

Ross, C. F., K. M. Weller, R. B. Blue, and J. P. Reganold. 2009. Difference testing of Merlot produced from biodynamically and organically grown wine grapes. *Journal of Wine Research* 20, 85–94.

Schindelbeck, R. R., and H. M. van Es. 2011. Understanding and managing your soil using the Cornell Soil Health Test. In *Proceedings of the 14th Australian Wine Industry Technical Conference, July 3–8,* 226–228. Adelaide: Australian Society of Viticulture and Oenology.

Shackel, K. 2006. Water relations of woody perennial plant species. In *Terroirs viticoles 2006,* eds. C. van Leeuwen and J. Fanet, 54–63. Bordeaux: Vigne et Vin Publications Internationales.

Smart, D. R., E. Schwass, A. Lakso, and L. Morano. 2006. Grapevine rooting patterns: A comprehensive analysis and a review. *American Journal of Enology and Viticulture* 57, 89–104.

Smart, R. E. 2001. Where to plant and what to plant. *Australian and New Zealand Wine Industry Journal* 16 (4), 48–50.

Smart, R. E., and P. R. Dry. 1980. A climatic classification for Australian viticultural regions. *Australian Grapegrower and Winemaker* 196, 8–16.

Smart, R., and M. Robinson. 1991. *Sunlight into wine: A handbook for winegrape canopy management*. Adelaide: Winetitles.

Smith, T. 2007. Terroirs of the Barossa. *Daily Wine News*, August 23 (www.winebiz.com.au). Adelaide: Winetitles.

Soil Survey Division Staff. 1993. *Soil survey manual*. 3rd ed. United States Department of Agriculture Handbook No. 18. Washington, DC: U.S. Government Printing Office.

Soil Survey Staff. 1999. *Soil taxonomy. A basic classification for making and interpreting soil surveys*. 2nd ed. United States Department of Agriculture, Natural Resources Conservation Service Handbook No. 436. Washington, DC: U.S. Government Printing Office.

Swinchatt, J., and D. G. Howell. 2004. *The winemaker's dance: Exploring terroir in the Napa Valley*. Berkeley: University of California Press.

Tesic, D., D. J. Woolley, E. W. Hewett, and D. J. Martin. 2002. Environmental effects on cv Cabernet Sauvignon *(Vitis vinifera L.)* grown in Hawke's Bay, New Zealand. II: Development of a site index. *Australian Journal of Grape and Wine Research* 8, 27–35.

Thorn, R. G. 2000. Soil fungi. In *Handbook of soil science*, ed. M. E. Sumner, C-22–C-37. Boca Raton, FL: CRC Press.

Van Leeuwen, C. 2010. Terroir: The effect of the physical environment on vine growth, grape ripening and wine sensory characteristics. In *Managing wine quality, Volume 1: Viticulture and wine quality*, ed. A. G. Reynolds, 365–444. Cambridge: Woodhead Publishing.

Walker, G. E., and G. R. Stirling. 2008. Plant-parasitic nematodes in Australian viticulture: Key pests, current management practices and opportunities for future improvements. *Australasian Plant Pathology* 37, 268–278.

Weckert, M., L. Rahman, and M. Alonso. 2009. Cover crops and composts in vineyards—Opportunities for improving soil health. In *Vineyard soil health—Unearthing fact from fiction*. Proceedings of the Australian Society of Viticulture and Oenology Seminar, eds. N. Blieschke, J. Henderson, E. Heyes, R. Johnstone, M. Krstic, R. Learmount, G. McCorkelle, and D. Robinson, 18–20. Adelaide: Australian Society of Viticulture and Oenology.

White, R. E. 2003. *Soils for fine wines*. New York: Oxford University Press.

White, R. E. 2006. *Principles and practice of soil science*. 4th ed. Oxford: Blackwell Publishing.

White, R. E. 2013. Has soil drying contributed to earlier grape ripening in wine regions of southern Australia? *Australian Journal of Grape and Wine Research* 19, 123–127.

Whiting, J. R. 2003. *Selection of grapevine rootstocks and clones*. Melbourne, State of Victoria: Department of Primary Industries.

Whiting, J. R. 2004. Grapevine rootstocks. In *Viticulture. Volume 1: Resources*. 2nd ed., eds. P. R. Dry and B. G. Coombe, 167–188. Adelaide: Winetitles.

Wilson, J. 2008. A grave waste, or fertiliser? *Sydney Morning Herald*, August 6 (smh.com. au). Sydney: News Corporation Ltd.

Wine Institute of California. 2005–2008. Potential benefits of sustainable winegrowing practices. (www.wineinstitute.org/initiatives/sustainablewinegrowing/benefits). San Francisco: Wine Institute of California.

Winter, E., S. Lowe, and L. Campbell. 2013. Biochar applications in a King Valley vineyard. *Australian and New Zealand Grapegrower and Winemaker* 597, 26–32.

Wratten, S. 2009. Quoted by Max Allen in *Daily Wine News*, July 17 (www.winebiz.com. au). Adelaide: Winetitles.

Zhang, X., R. R. Walker, R. M. Stevens, and L. D. Prior 2002. Yield-salinity relationships of different grapevine (*Vitis vinifera* L.) scion-rootstock combinations. *Australian Journal of Grape and Wine Research* 8, 150–156.

Web Sites

http://rodaleinstitute.org/ (accessed June 2013)
www.ams.usda.gov (accessed September 2013)
www.clw.csiro.au/publications/projects/projects20.pdf (accessed January 2014)
www.daff.gov.au/agriculture-food/food/organic-biodynamic (accessed July 2014)
www.ecocert.com (accessed September 2013)
www.ifoam.org (accessed September 2013)
www.ipw.co.za (accessed January 2014)
www.ithaka-journal.net (accessed September 2013)
www.natrakelp.com.au (accessed September 2013)
www.nzwine.com (accessed January 2014)
www.ofa.com.au (accessed September 2013)
www.redwhiteandgreen.com.au (accessed September 2013)
www.scew.gov.au/home (accessed September 2013)
www.vitiplus.ch (accessed January 2014)
www.wfa.org.au/entwineaustralia/ (accessed January 2014)
www.wineinstitute.org (accessed January 2014)

Index